SpringerBriefs in Applied Sciences and Technology

For further volumes:
http://www.springer.com/series/8884

Pratima Bajpai

Bleach Plant Effluents from the Pulp and Paper Industry

Springer

Pratima Bajpai
Thapar Centre for Industrial R&D
Patiala
Punjab
India

ISSN 2191-530X   ISSN 2191-5318   (electronic)
ISBN 978-3-319-00544-7   ISBN 978-3-319-00545-4   (eBook)
DOI 10.1007/978-3-319-00545-4
Springer Cham Heidelberg New York Dordrecht London

Library of Congress Control Number: 2013939564

Springer is part of Springer Science+Business Media (www.springer.com)

Preface

The pulp and paper industry is considered to be one of the most polluting in the world. The production process consists of two main steps: pulping and bleaching. These processes are very energy and water intensive.

Of the different wastewaters generated by the pulp and paper industry, bleach plant effluents are considered to be the most polluting. Pollutants such as chlorinated phenolics and dioxins are toxic, non-biodegradable, and have a tendency to contaminate food chains through bioaccumulation. Dioxins are known for their extreme toxicity and are believed to be carcinogenic. Bleach effluents are colored. They contain chlorinated and non-chlorinated products of lignin and extractives of wood. Because of the color, productivity of aquatic ecosystems gets affected when these effluents are discharged into the water bodies. Color also affects the downstream uses (municipal and industrial) of water. It makes water treatment difficult and costly. "The chlorinated organic compounds and the lignin derivatives of the bleach effluents are recalcitrant and get bioaccumulated in food chains of aquatic ecosystems. The low molecular weight fraction of bleach effluent contains potentially problematic (toxic) compounds. These have the ability to penetrate cell membranes and a tendency to bioaccumulate. Low molecular weight chlorinated organic compounds significantly affect the biology of aquatic ecosystems. Disappearance of benthic invertebrates, high incidence of fish diseases, and mutagenic effects on the aquatic fauna are some of the consequences of the disposal of bleach effluents into surface waters" Bajpai (2012). The growing public awareness of the fate of these pollutants and stringent regulations established by various governmental authorities such as provincial and federal agencies are forcing the industry to treat effluents to the required compliance level before discharging them into the environment. Many studies have been conducted so far on this sector regarding the impact of as well as the control of pollutants. This book describes the environmental impact of the blanch plant effluents, environmental regulations, and strategies used for reducing the generation of pollutants. Both internal process modification and external treatment are discussed and the performance of the effluent treatment measures in use are compared.

Reference

Bajpai P (2012) Biotechnology for pulp and paper processing. Springer Science+Business Media, New York, p 414

Contents

Chapter 1
Background

Some excerpts taken from Bajpai (2012a). Biotechnology for Pulp and Paper Processing with kind permission from Springer Science+Business Media.

The global pulp and paper industry is in physical terms one of the largest industries in the world. It is dominated by North American and Northern European countries and East Asian countries like Japan. Australasia and Latin America also have significant pulp and paper industries. Both India and China are expected to be key in the industry's growth over the next few years. World production of paper and paper board totals around 390 million tons. Growth is the most rapid in Asia, which accounts for almost 40 % of total world paper and paperboard production, while the European Union and North America account for about one-quarter each.

The pulp and paper industry plays an important role in a country's economic growth. It is highly capital intensive and has been periodically affected by overcapacity. It is traditionally known to be a large contributor to environmental pollution due its large consumption of energy and chemicals.

The pulp and paper industry is a water intensive industry and ranks third in the world, after the primary metals and the chemical industries, in terms of fresh water withdrawal.

The consumption varies with the type of paper being produced and can be as high as 60 m^3 per ton of paper produced, even with the most modern and efficient operational techniques. However, the practice of recycling a certain amount of water is commonplace, as this recovers some of the fibers which have escaped in the wastewater. Recycling is achieved by use of systems promoting recirculation of the process waters. Thus, the consumption of freshwater is reduced.

1.1 Pollution Problems of Pulp and Paper Industry

The pulping and bleaching steps generate most of the liquid, solid, and gaseous wastes (Table 1.1) (Smook 1992).

Pulping and bleaching operations are energy intensive and typically consume huge volumes of fresh water and large quantities of chemicals. A partial list of the various types of compounds found in spent liquors generated from pulping and

P. Bajpai, *Bleach Plant Effluents from the Pulp and Paper Industry*,
SpringerBriefs in Applied Sciences and Technology,
DOI: 10.1007/978-3-319-00545-4_1, © The Author(s) 2013

Table 1.1 Pollutants generated during chemical pulping and bleaching

Wood debarking and chipping, chip washing
Bark, wood processing residues, SS, BOD, color, resin acids
Chemical pulping, black liquor evaporation, and chemical recovery step
Total-reduced sulfur (hydrogen sulfide, methyl mercaptan, dimethyl sulfide, dimethyl disulfide),
    VOC
Wood chip digestion, spent pulping liquor evaporator condensates
High BOD, color, may contain reduced sulfur compounds, resin acids
Pulp screening, thickening, and cleaning operations
Large volume of waters with SS, BOD, color
Smelt dissolution, clarification to generate green liquor
Green liquor dregs
Recausticizing of green liquor, clarification to generate white liquor
Lime slaker grits
Chlorine bleaching of pulp
BOD, color, chlorinated organics, resin acids
Wastewater treatment
Primary and secondary sludge, chemical sludge
Scrubbing (flue gases)
Scrubber sludge
Recovery furnaces and boilers
Fine and coarse particulates, nitrogen oxides, SO2, ash

Based on Smook (1992)

bleaching steps is shown in Table 1.2 (Kringstad and Lindstrom 1984; U.S. EPA (1997).

The effluents generated by the mills are associated with many problems (Sumathi and Hung 2006). For example, dark brown coloration of the water bodies receiving effluents results in reduced penetration of light, thereby affecting benthic growth and habitat. Lignin and its degradation products are responsible for numerous problems:

- Discoloration of bodies of water into which effluents drain.
- High content of organic matter, which contributes to BOD and depletion of dissolved oxygen in the receiving ecosystems.
- Persistent, bioaccumulative, and toxic pollutants.
- Adsorbable organic halide load in the receiving ecosystems.
- Long-distance transport of organic halides such as chloroguaiacols thereby contaminating remote seas and lakes, and
- Cross-media pollutant transfer through volatilization of compounds and absorption of chlorinated organics to wastewater particulates and sludge.

Significant solid wastes from pulp and paper mills include bark, reject fibers, wastewater treatment plant sludge, scrubber sludge, lime mud, green liquor dregs, and boiler and furnace ash. The bulk of the solid waste is generated during wastewater treatment. Sludge disposal is a serious environmental problem due to the

Table 1.2 A partial list of the compounds found in spent liquors generated from pulping and bleaching steps

Acidic	Phenolic	Neutral	Miscellaneous
Wood extractives	*Phenolic*	Hemicelluloses	*Dioxins*
		Methanol	
Fatty acid	Monochlorophenols	Chlorinated	2,3,7,8-tetrachlorod-
		acetones	ibenzodioxin
	Dichlorophenols	Chloroform	
Formic acid (S)	Trichlorophenols	Dichloromethane	2,3,7,8-tetrachlorod-
Acetic acid (S)	Tetrachlorophenol	Trichloroethene	ibenzofuran
Palmitic acid (S)	Pentachlorophenol	Chloropropenal	
Heptadecanoic acid (S)		Chlorofuranone	*Wood derivatives*
Stearic acid (S)	*Guaiacolic*	1,1-dichloromethyl-	Monoterpenes
		sulfone	
Arachidic acid (S)			
Tricosanoic acid (S)	Dichloroguaiacols	Aldehydes	Sesquiterpenes
Lignoceric (S)	Trichloroguaiacols	Ketones	Diterpenes: Pimarol
Oleic (US)	Tetrachloroguaiacol	Chlorinated sulfur	Abienol
Linolenic acid (US)		Reduced sulfur	
		compounds	
Behenic acid (S)	*Catecholic*		*Juvabiones*
Resin acid	Dichlorocatechols		Juvabiol
	Trichlorocatechols		Juvabione
Abietic acid			
Dehydroabietic acid	*Syringic*		*Lignin derivatives*
Mono and dichloro	Trichlorosyringol		Eugenol
dehydrabietic acids			
Hydroxylated	Chlorosyringaldehyde		Isoeugenol
dehydrabietic acid			
Levopimaric acid			Stilbene
Pimaric acid			Tannins
Sandracopimaric acid			Flavonoids
Lignin/carbohydrate			
derived			
Hydroxy Glyceric acid			
Dibasic			
Oxalic acid			
Malonic acid			
Succinic acid			
Malic acid			
Phenolic acid			
Monohydroxy benzoic acid			
Dihydroxy benzoic acid			
Guaiacolic acid			
Syringic acid			

Based on Kringstad and Lindstrom (1984) and U.S. EPA (1997)

partitioning of chlorinated organics from effluents to solids. The major air emissions are fine and coarse particulates from recovery furnaces and burners, sulfur oxides from sulfite mills, reduced sulfur gases and associated odor problems from Kraft pulping and chemical recovery operations, volatile organic compounds from wood chip digestion, spent liquor evaporation and bleaching, nitrogen oxides and SOx from combustion processes. Volatile organics include carbon disulfide, methanol, methyl ethyl ketone, phenols, terpenes, acetone, alcohols, chloroform, chloromethane, and trichloroethane.

The extent of pollution and toxicity depends very much upon the raw material used, pulping method, and the bleaching process used by the pulp and paper mills. The pollution load from hardwood is lower than softwood. On the other hand, the spent liquor generated from pulping of nonwood fiber has a high silica content. Volumes of wastewater discharged varies significantly depending on the raw material used, manufacturing process, and the size of the mill (Rintala et al. 1999). Thus, the variability of effluent characteristics and volume from one mill to another emphasizes the requirement for a variety of pollution prevention and treatment technologies, tailored for a specific mill.

Increasing awareness of environmental consequences of bleach effluent has led to stringent environmental regulations (Bajpai 2012a). Most nations have imposed limits on adsorbable organic halides (AOX) of the effluents. In some nations, limits have also been set on individual chlorinated organic compounds of bleach effluents. In response to environmental concerns and environmental regulations, the pulp and paper industry has reacted by making process modifications based on existing and new proven technologies. For bleached Kraft mills, a number of alternative technologies are available (Bajpai 2012b). It is possible to select a combination that can meet the present or future effluent discharge limits. Initially, the effluent requirements varied from country to country for reasons of differing national priorities and this has led to diverse range of technological responses. However, as a result of concern regarding dioxins and polychlorinated organic materials, and as a result of more stringent regulations, the industry tends to evaluate and avail itself of the multitude of wide ranging options available.

References

Bajpai P (2012a) Biotechnology for pulp and paper processing. Springer Science+Business Media, p 414
Bajpai P (2012b) Environmentally benign approaches for pulp bleaching. 2nd edn, Elsevier Science B.V, p 416
Kringstad KP, Lindstrom K (1984) Spent liquors from pulp bleaching (critical review). Environ Sci Technol 18:236A–247A
Rintala JA, Jain VK, Kettunen RH (1999) Comparative status of the world-wide commercially available anaerobic technologies adopted for biomethanation of pulp and paper mill effluents. In: 4th international exhibition and conference on pulp and paper industry, PAPEREX-99, New Delhi, India, 14–16 Dec 1999
Smook GA (1992) Handbook for pulp and paper technologists, 2nd edn. Angus Wilde Publications, Vancouver 419

Sumathi S, Hung YT (2006) Treatment of pulp and paper mill wastes. In: Wang LK, Hung YT, Lo HH, Yapijakis C (eds) Waste treatment in the process industries. Taylor & Francis, USA, pp 453–497. ISBN 0-8493-7233-X

U. S. EPA (1997) Technical support document for best management practises for spent liquor management, spill prevention and control. USEPA, Washington, DC

Chapter 2
Pulp and Paper Making Process

Some excerpts taken from Bajpai (2012a). Biotechnology for Pulp and Paper Processing with kind permission from Springer Science+Business Media.

The paper manufacturing process has several stages—Raw material preparation and handling, Pulp manufacturing, Pulp Washing, and Screening, Chemical recovery, Bleaching, and finally Stock Preparation and Papermaking.

2.1 Pulp Making Process

Manufacturing of pulp starts with raw material preparation (Smook 1992; Biermann 1996). This includes debarking (when wood is used as raw material), chipping, and other processes such as depithing (for example, when bagasse is used as the raw material). Cellulosic pulp is manufactured from the raw materials, using chemical and mechanical means (Bajpai 2012a). The manufacture of pulp for paper and board employs mechanical (including thermomechanical), chemi-mechanical, and chemical methods. Each pulping process has its advantages and disadvantages (Smook 1992; Biermann 1996). The major advantage of mechanical pulping is its high yield of fibers—up to 90 %. Chemical pulping yields approximately 50 % but offers higher strength properties.

Mechanical pulping separates fibers from each other by mechanical energy applied to the wood matrix causing the bonds between the fibers to break gradually and fiber bundles, single fibers, and fiber fragments to be released. It is the mixture of fibers and fiber fragments that gives mechanical pulp its favorable printing properties. In mechanical pulping, the objective is to maintain the main part of the lignin in order to achieve high yield with acceptable strength properties and brightness. Mechanical pulps have a low resistance to ageing which results in a tendency to discolor. Mechanical pulps are weaker than chemical pulps, but cheaper to produce and are generally obtained in the yield range of 85–95 %. Currently, mechanical pulps account for 20 % of all virgin fiber material. The main processes are stone groundwood pulping (SGW), pressure groundwood pulping (PGW), thermo-mechanical pulping (TMP), or chemi-thermo-mechanical pulping (CTMP).

P. Bajpai, *Bleach Plant Effluents from the Pulp and Paper Industry*,
SpringerBriefs in Applied Sciences and Technology,
DOI: 10.1007/978-3-319-00545-4_2, © The Author(s) 2013

Chemical pulping is used for most paper produced commercially in the world (Smook 1992; Biermann 1996). Chemical pulps are made by cooking the raw materials, using the Kraft (sulfate) and sulfite processes. The Kraft process is the dominant chemical pulping process. In the Kraft process the active cooking chemicals are sodium hydroxide and sodium sulphide. The Kraft process is applicable to all types of wood but its chemistry carries with it the inherent potential for malodorous compounds. Kraft pulp possesses superior pulp strength properties in comparison to sulfite pulp. Kraft processes produce a variety of pulps used mainly for packaging and high-strength papers and board.

Chemical recovery is an essential part of the pulp production process (Tran 2007; Vakkilainen 2000; Bajpai 2008; Biermann 1996). Half of the wood raw material is utilized as chemical pulp fiber. The other half is utilized as fuel for electricity and heat generation. In fact, a pulp mill has two main lines. Wood is turned into pulp on the fiber line. Energy is produced on the chemical recovery line from the wood material cooked in the liquor; the cooking chemicals are recovered for reuse. In the chemical recovery line, black liquor is evaporated and combusted in a recovery boiler, and the energy content of the dissolved wood material is recovered as steam and electricity. The chemical pulping process generates more energy than it uses. A pulp mill generates energy for its own use and energy to sell.

The sulfite process uses different chemicals to attack and remove lignin. The sulfite process is characterized by its high flexibility compared to the Kraft process. In principle, the entire pH range can be used for sulfite pulping by changing the dosage and composition of the chemicals (Smook 1992; Biermann 1996). The use of sulfite pulping permits the production of many different types and qualities of pulps for a broad range of applications.

After pulp production, pulp is processed in wide variety of ways to remove impurities, and any residual cooking liquor is recycled via the process. Some pulp processing steps that remove pulp impurities are screening, defibering, and deknotting. Residual spent cooking liquor from chemical pulping is washed from the pulp using brown stock washers for Kraft and red stock washers for sulfite. Efficient washing is critical to maximize return of cooking liquor and to minimize carryover of cooking liquor into the bleach plant.

Mechanical pulp can be used without bleaching to make printing papers for applications in which low brightness is acceptable—primarily, newsprint. However, for most printing, for copying, and for some packaging grades, the pulp has to be bleached (Smook 1992). For mechanical pulps, most of the original lignin in the raw pulp is retained but is bleached with peroxides and hydrosulfites. In the case of chemical pulps, the objective of bleaching is to remove the small fraction of the lignin remaining after cooking (Smook 1992; Reeve 1996a, b). Oxygen, hydrogen peroxide, ozone, peracetic acid, sodium hypochlorite, chlorine dioxide, chlorine, and other chemicals are used to transform lignin into an alkali-soluble form (Reeve 1989). An alkali is necessary in the bleaching process to extract the alkali-soluble form of lignin.

Pulp is washed with water in the bleaching process. In modern mills, oxygen is normally used in the first stage of bleaching (Bajpai 2012b). The trend is to avoid the use of any kind of chlorine chemicals and employ "total chlorine-free" (TCF) bleaching. TCF processes allow the bleaching effluents to be fed to the recovery boiler for steam generation; the steam is then used to generate electricity thereby reducing the amount of pollutants discharged. Elemental chlorine-free (ECF) processes, which use chlorine dioxide, are required for bleaching certain grades of pulp. The use of elemental chlorine for bleaching is not recommended. Only ECF processes are acceptable, and, from an environmental perspective, TCF processes are preferred. The soluble organic substances removed from the pulp in bleaching stages that use chlorine or chlorine compounds, as well as the substances removed in the subsequent alkaline stages, are chlorinated. Some of these chlorinated organic substances are toxic.

2.2 Stock Preparation and Paper Making Process

Stock preparation is conducted to convert raw stock into finished stock for the paper machine. The pulp is prepared for the paper machine by blending different pulps, dilution, and the addition of chemicals. The raw stocks used are the various types of chemical pulp, mechanical pulp, and recovered paper, and their mixtures. The quality of the finished stock essentially determines the properties of the paper produced. Stock preparation consists of several steps that are adapted to one another: fiber disintegration, cleaning, fiber modification, and storage and mixing. These systems differ considerably depending on the raw stock used and on the quality of furnish required. For instance, in the case of pulp being pumped directly from the pulp mill, the slushing and deflaking stages are omitted. The operations practiced in the paper mills are: dispersion, beating/refining, metering, and blending of fiber and additives.

Pulpers are used to disperse dry pulp into water to form a slurry. Refining is one of the most important operations when preparing papermaking fibers (Baker 2000). The term beating is applied to the batch treatment of stock in a Hollander beater or one of its modifications. The term refining is used when the pulps are passed continuously through one or more refiners, whether in series or in parallel. Refining develops different fiber properties in different ways for specific grades of paper. Usually, it aims to develop the bonding ability of the fibers without reducing their individual strength by damaging them too much, while minimizing the development of drainage resistance. So the refining process is determined by the properties required in the final paper.

The furnish can also be treated with chemical additives. These include resins to improve the wet strength of the paper, dyes, and pigments to affect the color of the sheet, fillers such as talc and clay to improve optical qualities and sizing agents to control penetration of liquids and to improve printing properties. After stock preparation, the next step is to form the slurry into the desired type of paper at the wet end of the paper machine.

The pulp is pumped into the head box of the paper machine at this point (Smook 1992; Biermann 1996). The slurry consists of approximately 99.5 % water and

0.5 % pulp fiber. The exit point for the slurry is the "slice" or head box opening. The fibrous mixture pours onto a traveling wire mesh in the Fourdrinier process, or onto a rotating cylinder in the cylinder machine (Biermann 1996). The Fourdrinier machine is essentially a table over which the wire moves. Greater quantities of slurry released from the head box result in thicker paper. As the wire moves along the machine path, water drains through the mesh. Fibers align in the direction of the wire travel and interlace to improve sheet formation. After the web forms on the wire, the remaining task of the paper machine is to remove additional water. Vacuum boxes located under the wire aid in this drainage.

The next stop for the paper is the pressing and drying section where additional dewatering occurs (Smook 1992; Biermann 1996). The newly created web enters the press section and then the dryers. The extent of water removal from the forming and press sections depends greatly on the design of the machine and the running speed. When the paper leaves the press section, the sheet usually has about 65 % moisture content. The paper web continues to thread its way through the steam heated dryers loosing moisture each step of the way. The process evaporates many tons of water.

Paper will sometimes undergo a sizing or coating process. The web in these cases continues into a second drying operation before entering the calendaring stacks that are part of the finishing operation. Moisture content should be about 4–6 % as predetermined by the mill. If the paper is too dry, it may become too brittle. About 90 % of the cost of removing water from the sheet occurs during the pressing and drying operations. Most of the cost is for the energy required for drying.

At the end of the paper machine, paper continues onto a reel for winding to the desired roll diameter. For grades of paper used in the manufacture of corrugated paperboard, the process is now complete. For those papers used for other purposes, finishing and converting operations will now occur, typically off line from the paper machine. These operations can include coating, calendaring, or super calendaring and winding.

Other operations may include cutting, sorting, counting, and packaging. For some products such as tissue and copy paper, the typical paper mill will conduct all of these operations. In most cases, however, the rolls are wrapped and readied for shipment to their final destination.

References

Bajpai P (2008) Chemical recovery in pulp and paper making. PIRA International, UK, p 166

Bajpai P (2012a) Biotechnology for pulp and paper processing. Springer Science+Business Media, New York, p 414

Bajpai P (2012b) Environmentally benign approaches for pulp bleaching, 2nd edn, Elsevier Science B.V, p 416

Baker CF (2000) Refining technology. In: Baker C (ed) Pira International, Leatherhead, p 197

Biermann CJ (1996) Wood and fiber fundamentals. In: Handbook of pulping and papermaking. Academic Press, San Diego, p 754

Reeve DW (1989). Bleaching chemicals. In: Kocurek MJ (ed) Pulp and paper manufacture. Alkaline Pulping, Joint Textbook Committee of the Paper Industry, vol 5, p 425

Reeve DW (1996a) Introduction to the principles and practice of pulp bleaching. In: Dence CW, Reeve DW (eds) Pulp bleaching: principles and practice. Tappi Press, Atlanta, Section 1, Chapter 1, p 1

Reeve DW (1996b) Pulp bleaching: principles and practice. In: Dence CW, Reeve DW (eds) Chlorine dioxide in bleaching stages. Tappi Press, Atlanta, Section 4, Chapter 8, p 379

Smook GA (1992) Handbook for pulp and paper technologists, 2nd edn. Angus Wilde Publications, Vancouver, p 419

Tran H (2007) Advances in the Kraft chemical recovery process. In: Source 3rd ICEP international colloquium on eucalyptus pulp, Belo Horizonte, Brazil, 4–7 Mar 2007, p 7

Vakkilainen EK (2000) Chemical recovery, papermaking science and technology book 6B. In: Gullichsen J, Paulapuro H (eds) Fapet Oy, Chapter 1, p 7

Chapter 3
Pulp Bleaching and Bleaching Effluents

*Some excerpts taken from Bajpai (2012). Biotechnology for
Pulp and Paper Processing with kind permission from Springer
Science+Business Media.*

3.1 Pulp Bleaching

In Kraft pulping, about 90–95 % of wood lignin gets solubilized during the
cooking process. The remaining 5–10 % of lignin is responsible for the brown
color of the Kraft pulp and unbleached paper. The objective of bleaching is to
remove the residual lignin from the pulp as selectively as possible, without degrading the pulp carbohydrate, especially cellulose, which would lead to a decrease in
strength of the pulp. Other objectives are to increase the brightness of the pulp so
that it can be used in paper products such as printing grade and tissue papers. For
chemical pulps an important benefit is the reduction of fiber bundles and shives as
well as the removal of bark fragments. This improves the cleanliness of the pulp.
Bleaching also eliminates the problem of yellowing of paper in light, as it removes
the residual lignin in the unbleached pulp. Resin and other extractives present in
unbleached chemical pulps are also removed during bleaching, and this improves
the absorbency, which is an important property for tissue paper grades. In the manufacture of pulp for reconstituted cellulose such as rayon and for cellulose derivatives such as cellulose acetate, all wood components other than cellulose must
be removed. In this situation, bleaching is an effective purification process for
removing hemicelluloses and wood extractives as well as lignin. To achieve some
of these product improvements, it is often necessary to bleach to high brightness.
Thus, high brightness may in fact be a secondary characteristic of the final product and not the primary benefit. It is therefore simplistic to suggest that bleaching
to lower brightness should be practiced based on the reasoning that not all products require high brightness. The papermaking properties of chemical pulps are
changed after bleaching (Voelker 1979).

Bleaching is carried out in a multi-stage process that alternates between delignification and extraction of dissolved material. Additional oxygen- or hydrogen
peroxide-based delignification may be desired to reinforce the extracting operation. Since its introduction at the turn of the century, chemical Kraft bleaching
has been refined into a stepwise progression of chemical reaction, evolving from

P. Bajpai, *Bleach Plant Effluents from the Pulp and Paper Industry*,
SpringerBriefs in Applied Sciences and Technology,
DOI: 10.1007/978-3-319-00545-4_3, © The Author(s) 2013

a single-stage hypochlorite (H) treatment to a multi-stage process, involving chlorine (CI_2), chlorine dioxide (CIO_2), hydrogen peroxide and ozone (O_3). Bleaching operations have continuously evolved since the conventional chlorination-extraction-hypo-chlorine dioxide-extraction- chlorine dioxide (CEHDED) sequence and now involve different combinations with or without chlorine containing chemicals.

The introduction of Cl_2 and ClO_2 in the 1930s and early 1940s, respectively, increased markedly the efficiency of the bleaching process (Rapson and Strumila 1979; Reeve 1996a, b). Being much more reactive and selective than hypochlorite, Cl_2 had less tendency to attack the cellulose and other carbohydrate components of wood, producing much higher pulp strength. Although it did not brighten the pulp as hypochlorite, it extensively degraded the lignin, allowing much of it to be washed out and removed with the spent liquor by subsequent alkaline extraction. The resulting brownish Kraft pulp eventually required additional bleaching stages to increase brightness, which led to the development of the multi-stage process. Chlorine dioxide, a more powerful brightening agent than hypochlorite, brought the Kraft process efficiency one step further. Between the 1970s and 1990s, a series of incremental and radical innovations increased again the efficiency of the process, while reducing its environmental impact (Reeve 1996b). Development of oxygen delignification, modified and extended cooking, and improved operation controls improved the economic efficiency of the process and led to significant reduction of wastewater (McDonough 1995). In addition, higher CIO_2 substitution, brought down significantly the generation and release of harmful chlorinated organic compounds. Earlier, it was believed that a 90° brightness could not be achieved without the use of chlorine and chlorine containing chemicals as bleaching agents. The implementation of modified cooking and oxygen-based delignification impacted the entire process by lowering the kappa number of the pulp prior to bleaching, thereby reducing further the amount of bleaching chemicals needed. Under tightening regulations and market demands for chlorine-free products, the industry eventually accelerated the implementation of ECF and TCF bleaching processes, by substituting oxygen-based chemicals for hypochlorite, CI_2 and CIO_2, although the timing and scale of these trends have varied between regions.

Single application of chemicals have a limited effect in brightness improvement or in delignification. Multi-stage application of bleaching chemicals can provide much greater benefits. A bleaching sequence for a chemical pulp consists of a number of stages. Each stage has a specific function (Reeve 1996a). The early part of a sequence removes the major portion of the residual lignin in the pulp. Unless this is done, a high brightness cannot be reached. In the later stages in the sequence, the so-called brightening stages, the chromophores in the pulp are eliminated and the brightness increases to a high level. Most bleaching chemicals are oxidizing agents that generate acidic groups in the residual lignin. If a bleaching stage is done under acidic conditions, it is followed by an alkaline extraction to remove the water-insoluble acidic lignin products. Modern bleaching is done in a continuous sequence of process stages utilizing different chemicals and conditions

Table 3.1 Commonly applied chemical treatments

Oxygen (O_2) delignification (Reaction with O_2 at high pressure in alkaline medium)	O
Chlorine (Cl_2) (Reaction with elemental chlorine in acidic medium)	C
Chlorine dioxide (ClO_2) (Reaction with ClO_2 in acidic medium)	D
Mixtures of Cl_2 and ClO_2 (major component being first listed)	(C+D) & (D+C)
Hypochlorite (Reaction with hypochlorite in alkaline medium)	H
Alkaline extraction (Dissolution of reaction products with NaOH)	E
Alkaline extraction reinforced with O_2	(E_O)
Alkaline extraction with addition of hydrogen peroxide (H_2O_2)	(E_P)
Alkaline extraction with addition of H_2O_2 and reinforced with O_2	(E_{OP})
Treatment with a metal-chelating chemical	Q
Acid treatment	A
Ozone	Z
Alkaline hydrogen peroxide	P
Pressurized alkaline hydrogen peroxide treatment	(PO)
Peroxyacetic (peracetic) acid	Pa
Peroxymonosulfuric acid	Px

in each stage, usually with washing between stages. The commonly applied chemical treatments and their symbols are shown in Table 3.1.

Selectivity and cost are two important aspects of establishing the proper sequence of bleaching chemicals. Chlorine is less expensive than chlorine dioxide, so chlorine should be used earlier in the sequence where more lignin is present and more chemical is required (Reeve 1996a). Thus, for greater economy, the preferred sequence is CED instead of DEC. Also, chlorine is less selective than chlorine dioxide, so the former should not be used at the end of the sequence where the lignin content is low and the possibility of degrading carbohydrates is greater. Thus, to achieve greater selectivity, the preferred sequence is CED instead of DEC. Sulfite pulps and hardwood Kraft pulps are "easier bleaching" than softwood Kraft pulps. Both have lower lignin content, and in the case of sulfite pulps the lignin residues are partially sulfonated and more readily solubilized. Consequently, a somewhat simpler process can be used to achieve a comparable brightness level. For softwood Kraft pulps, a number of bleach sequences utilizing between four and six stages are commonly used to achieve "full-bleach" brightness levels of 89–91 %. ECF bleaching involves replacement of all the molecular chlorine in a bleaching sequence with chlorine dioxide. The term ECF bleaching is usually interpreted to mean bleaching with chlorine dioxide as the only chlorine-containing bleaching chemical. TCF bleaching uses chemicals that do not contain chlorine, such as oxygen, ozone, hydrogen peroxide, and peracids. The main bleaching chemicals can be divided into three categories based on their reactivity (Gierer 1990; Lachenal and Nguyen-Thi 1993). Table 3.2 presents this concept.

Each chlorine-containing chemical has an equivalent oxygen-based chemical. Ozone and chlorine are placed in the same category because they react with

Table 3.2 Classification of bleaching chemicals

Category		
Reaction with any phenolic group + double bond	Reaction with free phenolic group + double bond	Reaction with carbonyl groups
Cl_2	ClO_2	NaOCl
O_3	O_2	H_2O_2

Based on Lachenal and Nguyen-Thi (1993)

aromatic rings of both etherified and nonetherified phenolic structures in lignin and also with the double bonds. Ozone is found to be less selective than chlorine, as it attacks the carbohydrates in pulp. These chemicals are well suited for use in the first part of a sequence, as they are very efficient at degrading lignin. Chlorine dioxide and oxygen are grouped together because they both react primarily with free phenolic groups. They are not as effective as chlorine and ozone in degrading lignin. Chlorine dioxide is used extensively in the early stages of bleaching sequences as a replacement for chlorine even though it is slightly less effective. The classification is somewhat simplistic in this respect and, does not take into account that chlorine dioxide is reduced to hypochlorous acid and that oxygen is reduced to hydrogen peroxide during the bleaching reactions. Chlorine dioxide is also used as a brightening agent in the latter part of a sequence. Sodium hypochlorite and hydrogen peroxide react almost exclusively with carbonyl groups under normal conditions. This results in the brightening of pulp without appreciable delignification. The selection of a suitable bleaching sequence based on this classification has been discussed by Lachenal and Nguyen-Thi (1993). An efficient bleaching sequence should contain at least one chemical from each category.

3.2 Bleaching Effluents

During bleaching, the wood components, mainly lignin, get degraded, heavily modified, chlorinated, and finally, dissolved in the effluent. As a result, the effluent from the bleaching process is dark brown in color due to the presence of chromophoric polymeric lignin derivatives. The amount of chlorinated organics produced, during the pulp bleaching, varies with wood species, kappa number of the pulp, bleaching sequence, and conditions employed. Typically, color, BOD_5, COD, and AOX in the effluent, from the bleaching of softwood Kraft pulp by conventional sequence, are in the range of 150–200, 8–17, 50–70, and 3–5 kg/ton of pulp bleached, respectively (Springer 1993). The pollution loads from a hardwood Kraft pulp bleach plant are, generally, lower than those from a softwood pulp bleaching plant.

Of the total chlorine used in the bleach plant, about 90 % forms common salt, and 10 % or less gets bound to the organic material removed from the pulp. This

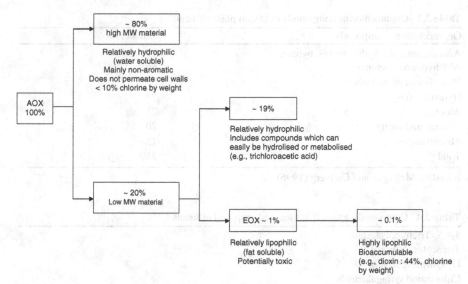

Fig. 3.1 AOX in the effluent from conventionally pulped and bleached Kraft pulp. Based on Gergov et al. 1988

organically bound chlorine is termed AOX. A physico-chemical classification of this chlorinated organic material, present in spent liquors from conventionally pulped and bleached softwood Kraft pulp is shown in Fig. 3.1 (Axegard et al. 1993; Gergov et al. 1988; Lindstrom and Mohamed 1988). The figure illustrates that as much as 80 % or more of the organically bound chlorine corresponds to high molecular weight (MW > 1,000) chlorinated lignin material, commonly referred to as chlorolignin. The exact chemical nature of chlorolignin is not clearly understood, but it is assumed to include mainly chlorine substituted polycarboxylic acid polymers, originating from the oxidative degradation of residual lignin, and devoid of aromatic structure.

About 20 % of the organically bound chlorine corresponds to relatively low molecular weight material. This fraction is expected to contain potentially problematic compounds due to their ability to penetrate cell membranes or their tendency to bioaccumulate in the fatty tissues of higher organisms. Some of the major components of this low molecular weight fraction have been found to consist of relatively water soluble substances, such as chlorinated acetic acids or chlorinated acetone (Gergov et al. 1988; Lindstrom and Mohamed 1988) which are easily broken down before or during biotreatment and are, therefore, of little environmental significance.

The fraction of AOX which is extractable by a nonpolar organic solvent and referred to as EOX, accounts for about 1–3 % of TOCl (Bajpai 2012). This fraction contains lipophilic neutral organic compounds primarily of low molecular weight and, therefore, of greater environmental significance than the remaining 99 % of the AOX material. About 456 different compounds have been identified in

Table 3.3 Organochlorine compounds in bleach plant effluents

Organochlorine compounds	Numbers
Acids, esters, aldehydes, furans, pyrenes	77
Aldehydes and ketones	66
Phenols and phenol ethers	52
Hydrocarbons	75
Alcohols	25
Dioxins and furans	20
Miscellaneous	15
Total	330

Based on McKague and Carlberg (1996)

Table 3.4 Compounds detected in bleached pulp mill effluents

3,4,5,-Trichlorocatechol
Tetrachlorocatechol
Dehydroabietic acid
Chlorinated syringaldehyde
3,4,5 Trichlorophenol
Trichloroguaicol
Tetrachloroguaicol
2,4,6 Trichlorophenol 2-Butanone
Toluene
Linoleic acid
Pentadecane
2-Butanone
Methanol
Dehydroabietic acid
Toluene

the effluents from conventional bleach plants. About 330 of those are organochlorine compounds, which include chlorinated phenolics, dioxins, hydrocarbons, and resin acids (Table 3.3) (Mckague and Carlberg 1996). The most common chlorinated phenolics in bleached Kraft pulp mill effluents are tri- and tetra chloroguaiacols (Table 3.4) (Liebergott et al. 1990).

References

Axegard P, Berry RM, Gellerstedt G, Lindblad PO, Luthie CE, Popke I, Voss RH, Wrist PE (1993) The effects of recent changes in bleached softwood Kraft mill technology on organochlorine emissions: an international perspective. Bleaching (Jameel H ed) Tappi Press, Atlanta Georgia, vol 2, pp 759–770
Bajpai P (2012) Biotechnology for pulp and paper processing. Springer Science+Business Media, New York, p 414
Gergov M, Priha M, Talka E, Valtilla O, Kangas A, Kukkonen K (1988) Chlorinated organic compounds in effluent treatment at Kraft mills. Tappi J 71(12):175–184

Gierer J (1990) Basic principles of bleaching. 1. Cationic and radical processes. Holzforschung 44(5):387–394

Lachenal D, Nguyen-Thi NB (1993) Rationalization of chlorine-free bleaching. In: Proceedings 7th international symposium on wood and pulping chem. Beijing, PR China, vol 1, p 166

Liebergott N, van Lierop B, Kovacs T, Nolin A (1990) Comparison of the order of addition of chlorine and chlorine dioxide in the chlorination stage. Tappi J 73:207–213

Lindstrom K, Mohamed M (1988) Selective removal of chlorinated organics from Kraft mill effluents in aerated lagoons. Nordic Pulp and Paper Res J 3:26–33

McDonough TJ (1995) Recent advances in bleached chemical pulp manufacturing technology: part 1. Tappi J 78(3):55–62

Mckague AB, Carlberg G (1996) Pulp bleaching and the environment. In: Dence CW, Reeve DW (eds) Pulp bleaching: principles and practice. Tappi Press, Atlanta, Ga, pp 746–820

Rapson WH, Strumila GB (1979) The bleaching of pulp. In: Singh RP (ed) Chlorine dioxide bleaching, 3rd edn, Tappi Press, Chapter 6, p 113

Reeve DW (1996a) Introduction to the principles and practice of pulp bleaching. In: Dence CW, Reeve DW (eds) Pulp bleaching: principles and practice. Tappi Press, Atlanta, p 1, Section 1, Chapter 1

Reeve DW (1996b) Pulp bleaching: principles and practice. In: Dence CW, Reeve DW (eds) Chlorine dioxide in bleaching stages. Tappi Press, Atlanta, Section 4, Chapter 8, p 379

Springer AM (1993) Overview of water pollutants and their impact: pulp and paper industry. In: Springer AM (ed) Industrial environmental control: pulp and paper industry. Tappi Press. Atlanta, Ga., U.S.A., pp 7–34

Voelker MH (1979) The bleaching of pulp, 3rd edn. In: Singh RP (ed) Tappi Press, Atlanta, Ga, p 337

Chapter 4
Environmental Impact of Bleach Effluents

Some excerpts taken from Bajpai (2012). Biotechnology for Pulp and Paper Processing with kind permission from Springer Science+Business Media.

Bleach Kraft mill effluent is a complex mixture of chlorinated and non-chlorinated products of lignin and/or extractives of wood that imparts dark color to the effluent (Bajpai 1996; Bajpai 2012). Colored effluent may result in the following detrimental effects upon the receiving water body:

- Color, derived from lignin, is an indicator of the presence of potentially inhibiting compounds.
- Color reduces the visual appeal and recreational value of the water.
- Color-imparting substances complex with metal ions, such as iron, or copper, and form tar like residues. These residues may have direct inhibitory effects on some of the lower organisms in the food chain.
- Chromophoric groups exert long-term BOD (20–100 days) that cannot be measured in terms of 5-day BOD.
- Color affects downstream municipal and industrial water uses, and increases the cost and difficulty of pre-treatment for industrial processes.
- Color retards sunlight transmission, thus, reducing the productivity of the aquatic community by interfering with the photosynthesis.

Bleached Kraft mill effluent may have a noticeable effect on the biological quality of the receiving water. Disappearance of benthic invertebrates, such as mussels, and high incidence of fish diseases are some of the effects (Sundelin 1988; Sodergren et al. 1993). Bleached Kraft and bleached sulphite mill effluents have been demonstrated to impair the functions of liver, enzyme systems, and metabolic cycles in exposed fish. Furthermore, such exposures have been demonstrated to increase the incidence of spinal deformities and to reduce gonad development.

The majority of the organically bound chlorine (80 %) is believed to be heterogeneous material of relatively high molecular weight compounds. These compounds, apparently, contribute little to the effluent BOD and acute toxicity. Their major contribution is toward color, COD, and chronic toxicity. Natural ecological processes, such as sedimentation, biodegradation, and bioaccumulation are apparently correlated to the molecular size and hydrophobicity of the compounds. Highly polar and high molecular mass constituents are responsible for the toxicity

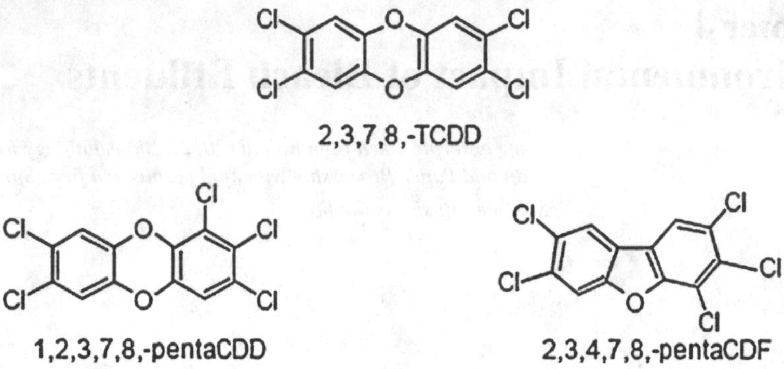

2,3,7,8,-TCDD

1,2,3,7,8,-pentaCDD   2,3,4,7,8,-pentaCDF

Fig. 4.1 Structures of the most toxic forms of dioxin and furan molecules, Based on Rappe (1995), Gavrilescu (2006)

of the bleach effluents during early life stages of marine animals and plants (Higachi et al. 1992). Chlorocymenes and chlorocymenenes, of the bleach effluent, have been reported to accumulate in fish and mussels (Suntio et al. 1988).

Chlorinated dioxins, which are present in very low concentrations in the bleach plant effluent (usually in ppt levels), account for a 10 billionth of the total AOX discharged. About 210 dioxins, belonging to the two families, namely, PCDDs, and PCDFs, have been reported in bleach effluents. The compounds 2,3,7,8-TCDF, and 2,3,7,8-TCDD are important from the toxicological point of view. These two chemicals are known to be highly toxic, carcinogenic, and bio-accumulable. The structures of the most toxic forms of dioxin and furan molecules are shown in Fig. 4.1 (Rappe and Wagman 1995). Dioxins are almost insoluble in water. They tend to enter the food chains and accumulate in high concentrations in predators, such as fish eating birds (McCubbin 1989; McCubbin et al. 1990). Adverse effects of dioxins have been observed in almost all species tested. According to an environmental protection agency (EPA) report (Anonymous 1994), human beings lie somewhere in the middle of the sensitivity range (from extremely responsive to extremely resistant) for dioxins. Even in trace amounts, dioxins may cause a wide range of adverse health conditions, such as disruption of regulatory hormones, reproductive and immune system disorders, abnormal fetal development (Bajpai and Bajpai 1996). Some of the toxic effects of selected organochlorines are given in Table 4.1.

4.1 Environmental Fate and Effects

The toxic effect on various fish species due to exposure to pulp and paper mill effluents has been reviewed (Pokhrel and Viraraghavan 2004). Respiratory stress, mixed function oxygenase (MFO) activity, mutagenicity, liver damage, and

Table 4.1 Toxic effects of some AOX compounds

Chlorinated dibenzodioxins and dibenzofurans
Highly toxic. Acute exposures cause severe skin rash, changes in skin color, hyperpigmentation, polyneuropathies in arms and legs. They act as endocrine disrupting factors by interfering with the production, release, transport, metabolism, binding action, or elimination of natural hormones in the body. They may cause reproductive and immune system disorders and abnormal fetal development. In fish, they decrease growth rate, increase egg mortality, and produce histological changes in liver
Chlorophenols
2,4-Dichlorophenol (DCP), 2,4,5-Trichlorophenol (TCP), and PCP are Group 2B carcinogens. PCP is the most toxic chlorophenol. Chronic exposition results in liver and kidney damage, loss in weight, general fatigue, and low appetite. In fish, these compounds cause impaired function of liver, enzyme system, metabolic cycle, increase in the incidence of spinal deformities, and reduced gonad development
Chlorobenzenes
Exposure to 60 ppm is known to cause drowsiness, headache, eye irritation, and sore throat. Chronic exposure is known to cause adverse effects on lungs, renal degeneration, and porphyria. Hexachlorobenzene is carcinogenic in animal tests. Monochlorobenzene is known to cause multiple effects on central nervous system—headache, dizziness, cyanosis, hyperesthesia, and muscle spasms
Chlorocatechols
Strong mutagens
Chloroguaiacols
Tetra- and tri-chloroguaiacols are known to bioaccumulate in fish

Based on Savant et al. (2006)

genotoxic effects have all been reported by Erisction and Larsson (2000) as deleterious effects on fish of paper and pulp plant pollutants. Yen et al. (1996) observed sublethal effects on the aquatic organisms in the Dong Nai River in Vietnam due to the effluents discharged from a pulp and paper mill. But, there are also some opposing reports by other researchers. Serious concerns related to the surface plankton population change in Elengabeel's wetland ecosystem in India due to untreated paper mill effluent discharge into the system have been reported by Baruah (1997). No significant adverse effect in sediments and river biota or on fish attributable to the treated mill effluent has been observed by Felder et al. (1998). Kovacs et al. (2002) did not observe much evidence of depressed plasma steroids. They also did not observe an increase in MFO activity in fish associated with pulp mill effluent. Stepanova et al. (2000) found no clear evidence of mutagens in most of aquatic animals studied in Lake Baikal due to Baikalsk pulp and paper mill wastewater discharged into the lake. Wayland et al. (1998) did not find any effect on the tree swallow, which feeds on the insects downstream from pulp mills. Howe and Michael (1998) studied the effects of treated pulp mill effluent on irrigated soil in northern Arizona, which showed serious soil chemistry change. Dutta (1999) investigated the toxic effect of the treated paper mill effluent applied to a paddy field in Assam, India. Skipperud et al. (1998) found various trace metals in the pulp and paper mill effluents at low levels. King et al. (1999) observed high levels of Mn accumulation in the Crayfish exposed to the paper mill wastewater. Health impacts such as diarrhea, vomiting, headaches, nausea, and eye irritation on children and workers due to the pulp and paper mill wastewater discharged to the environment have been reported by Mandal and Bandana (1996). High carbon dioxide levels in pulp and paper mill effluents as a potential source of distress and toxicity to rainbow trout was observed by O'connor et al. (2000). Gupta (1997) reported high loads of organic pollutants derived from the paper mill wastewater in Tamilnadu, Punjab, and India. Singh et al. (1996) found a high level of coliform bacteria in the effluent. However, Archibald (2000) indicated that the presence of coliform bacteria in the pulp and paper effluent did not necessarily mean a health hazard to the environment unless pathogens were observed.

4.2   Environmental Issues

Environmental issues have emerged as crucial, strategic factors for western industrial enterprises. Environmental concern and awareness have grown considerably and have been globalized. There have been environmental campaigns on particular subjects such as industrial use of chlorine, dioxins from pulp bleaching, and old-growth forestry. The complexity and sophistication of environmental criteria have increased dramatically in the last three decades. Toxicity testing of bleach plant effluents previously was limited in Canada to the acute toxicity test for rainbow trout. Now, there is a very wide array of biological response tests. There has been a significant increase in the scope and restrictiveness of environmental regulations in the last three decades. In the 1970s, regulation of bioCOD, suspended solids, and the use of acute toxicity tests became common as the main concern of regulators was to avoid killing fish and prevent oxygen reduction of receiving waters (Rennel 1995; Nelson 1998). In the 1980s, conditions changed, especially after a Swedish study of the Norrsundet mill, which discharged into the Baltic Sea. Sodergren (1989) and Sodergren et al. (1988) observed considerable effects on fish which was attributed to the presence of organochlorine compounds in the effluent. Later, it was found that the results of the study were misinterpreted because the effect of pulping liquor spills as a source of toxicity was ignored (Owens 1991). However, the release of the report resulted in regulatory authorities applying stringent controls on the discharge of organochlorine compounds from bleached pulp mills which compelled the industry to develop technology to reduce or eliminate their formation. This resulted in the wider use of processes such as extended delignification, oxygen delignification, oxygen reinforcement in alkaline extraction stages, and the substitution of elemental chlorine for chlorine dioxide (McDonough 1995). Secondary treatment of effluent became more extensive, as this reduced the amount of organochlorine compounds. The organochlorines (measured as AOX) in pulp mill effluents decreased when the above-mentioned technologies were introduced. Testing of effluents of Swedish mills in model ecosystems led to the conclusion that it was not possible to predict the environmental impact of an effluent exclusively on the basis of its AOX content at levels below about 2 kg/ton of pulp (Haglind et al. 1993). Laboratory study of effluents from Canadian mills showed that there was no correlation between acute or sublethal toxicity and AOX levels below about 2.5 kg/ton (O'Connor et al. 1994). A study by Folke et al. (1996) showed that reduction of the AOX level in various mill effluents below 1.2 kg/ton did not result in a further reduction in toxicity (Table 4.2).

These results showed that this residual toxicity was due to components other than organochlorines. The environmental organization Greenpeace started a campaign in Germany in the late 1980s, that convinced many consumers in the German-speaking region of Europe that use of paper bleached with chlorine chemicals was unwanted (Rennel 1995). In Germany, the pulp is produced using the sulfite process. Sulfite pulps can be easily bleached without the use of chlorine chemicals. The high-quality pulp purchased from Scandinavia is made by the sulfate process, and at that time bleaching was done with chlorine chemicals. So, the pulp mills in that region were forced to adopt TCF bleaching to maintain their dominance of the bleached Kraft pulp market in Germany. The North American industry was not affected by the same need to change to TCF bleaching as it exports very little bleached pulp to Germany. However, in North America, the presence of dioxin in pulp mill effluents, in bleached pulp, and its accumulation in biota near pulp mills were the main environmental issues. In Canada, because of the high dioxin content of crabs, fishing grounds were closed (Muller and Halliburton 1990). The industry rapidly found methods to reduce or eliminate the formation of dioxins. It is well documented that increasing chlorine dioxide substitution eliminates 2,3,7,8-TCDD and 2,3,7,8-TCDF to non-detectable levels (Bajpai 1996). Increasing chlorine dioxide substitution also reduces the formation of polychlorinated phenolic compounds. At 100 % substitution, tri-, tetra-, and penta-chlorinated phenolic compounds were not detected (O'Connor et al. 1994). These results for dioxin, furan, and polychlorinated phenolic compounds were confirmed in a study of seven Canadian bleached Kraft pulp mills operating ECF (NCASI 1995). An ecological risk assessment of the organochlorine compounds produced with ECF bleaching reached the conclusion that the environmental risks from these compounds are insignificant at mills bleaching with chlorine dioxide, employing secondary treatment, and with receiving water dilutions typically found in North America (Solomon et al. 1996).

Table 4.2 Relationship between effluent toxicity and AOX	AOX (kg/ton pulp)	Toxicity (log TEF_{SA})
	0.35	8
	1.4	9
	2.0	35
	2.9	68
	3.5	72
	3.9	100

Based on data from Folke et al. (1996)

To achieve compliance with the Cluster rule, most of the mills in the USA are using 100 % chlorine dioxide substitution. The specific limit for AOX is 0.512 kg/ ton of pulp (annual average). Many studies have investigated the quality of bleaching effluents from ECF and TCF bleaching processes (Nelson 1998). The general conclusion from each of these studies is that the differences in toxicity in a wide variety of test species are not significant, and that secondary biological treatment removes toxicity from the effluents. The effluent from a modern bleached Kraft pulp mill is found to show a low level of toxicity when subjected to very sensitive tests, such as the induction of MFO enzymes in the livers of fish (Hodson et al. 1996). There is proof that chlorine compounds are not exclusively responsible for the induction, as it has been shown that an important source of MFO inducers in bleached Kraft pulp mill effluents was spent pulping liquor (Martel et al. 1994; Schnell et al. 1993). The chemical compounds causing the induction have not been identified, but wood extractives are supposed to be involved. Hewitt et al. (1996) have found that in Canada, MFO inducers were not removed by secondary treatment of the effluent. The residual low toxicity of effluents is one reason for the view that in the long term the only environmentally acceptable pulp mills may be those which have completely closed-cycle pulping and bleaching (Albert 1996). A number of companies are moving toward closed-cycle bleaching (Johansson et al. 1995). However, this approach must be used with caution, as it can transfer pollutants from the liquid effluent to the air and solid waste.

4.3 Environmental Regulations

Traditional concerns over oxygen deficiency, fiber deposition, and the health of the fish community in receiving waters, have led to the setting of discharge limits for so-called, conventional parameters, such as BOD, COD, TSS, and acute toxicity to fish (Table 4.3). Now, there has been a shift in emphasis toward concerns about the discharge of persistent and bio-accumulable compounds which have potential, at lower than lethal concentrations, for chronic (long-term) adverse biological

Table 4.3 Discharge limits for some conventional parameters

	Toxicity	BOD (kg/ton)	COD (kg/ton)	TSS[a] (kg/ton)
USA	Chronic and acute[b]	5.5–12.5	–	11.9–20.1
Finland	–	6.8–3.4	65	5–15
Sweden	–	7.5–17	39–107	0.3–5.8
Canada	$LC_{50} \geq 100$ %	7.5–1.5	–	11.25

[a]Because filters of different coarseness are used for TSS by various countries, the values in this column are not comparable
[b]Varies by province or state
Based on Axegard et al. (1993)

Table 4.4 AOX discharge limits in various countries

Country	Value (kg/ton)	Remark
Australia	1.0	For new mills
Austria	0.5–1.0	
Canada		
Alberta	0.29	
British Columbia	0–0.5	
Ontario	1.5	
Quebec	0.5–1.5	Lower limits for hardwood pulps
Finland	1.0–2.0	
Germany	1.0	
India	1.0	
Japan	1.5	
Norway	1.0–2.0	Lower limits for hardwood pulps
Sweden	0.3–0.5	Lower limits for hardwood pulps
USA	0.623	Monthly average
	0.512	Annual average
	0.951	Daily maximum
	Non-detect	Dioxin and 12 chlorinated phenolic
	31.9 pg/l	TCDF (daily maximum)
	4.14 g/ton	Chloroform

effects. Such concerns have focused attention on mills producing chemical pulps bleached with chlorine compounds. This has led, in turn, to the introduction of regulatory measures ranging from limits of overall discharge of chlorinated organic materials, such as AOX, to specific polychlorinated organic compounds 2,3,7,8-TCDDs and 2,3,7,8-TCDFs.

The Canadian Environmental Protection Act prohibits release of final effluent that contains any measurable concentrations of 2,3,7,8-TCDDs (>15 ppq) and 2,3,7,8-TCDFs (>50 ppq) (Canada Gazette 1992). Many countries, presently, prescribe 1 kg/ton of pulp as the standard for AOX discharge. This limit is imposed on sulfite pulp mills in countries like Austria, Germany, and Norway, probably because control of AOX discharge is relatively easier in these mills using softwood as raw material. In certain regions of Sweden, even Kraft mills have limits as low as 0.3 kg/ton of pulp. Some decided or planned regulations are shown in Table 4.4. The permitted levels for AOX are likely to decrease to less than 1 kg/ton of pulp in next few years, in many more countries. According to a decision taken by Paris Convention for the Prevention of Marine Pollution (PARCOM), for land based sources and rivers, 1 kg/ton is the limit agreed for AOX since 1995. This limit applies to effluents from all types of chemically bleached pulps and has been accepted by many European countries (Belgium, Denmark, France, Germany, Ireland, Luxembourg, Netherlands, Norway, Portugal, Spain, Sweden, and U.K.). The so-called "cluster rule", first proposed in December 1993, was signed by the EPA on November 14, 1997, after some

Table 4.5 Limits for existing bleached Kraft (paper grade) and soda mills in USA

Parameter	Daily maximum	Monthly average
2,3,7,8-TCDD (pg/L)	<ML[a]	n.a.[b]
2,3,7,8-TCDF (pg/L)	31.9	n.a.
Chlorinated phenolic (μg/L)	<ML	n.a.
Chloroform (g/K kg)[c]	6.92	4.14
COD (kg/K kg)	Reserved	Reserved
AOX (kg/K kg)	0.951	0.623
Color, acetone, MEK, methylene chloride	No limit	No limit

[a]*ML* minimum level. The level at which the analytical signal gives recognizable signals and an acceptable calibration point
[b]*n.a.* not applicable
[c]1 K kg = 1 metric ton
Based on Vice and Carrol (1998)

Table 4.6 Regulated chlorophenols

Trichlorosyringol
2,4,5-Trichlorophenol
2,4,6-Trichlorophenol
3,4,5-Trichlorocatechol
3,4,5-Trichloroguaiacol
3,4,6-Trichlorocatechol
3,4,6-Trichloroguaiacol
4,5,6-Trichloroguaiacol
Tetrachlorocatechol
Tetrachloroguaiacol
2,3,4,6-Tetracholorophenol
Pentachlorophenol

Based on Vice and Carrol (1998)

modifications (Anonymous 1997). Limits proposed by the US-EPA, for various parameters, are shown in Tables 4.5 and 4.6.

References

Albert RJ (1996) Current status of TCF bleached kraft pulp mills and future trends. In: Proceedings international non-chlorine bleaching conference, Orlando, Florida, paper 12-3, p 61

Anonymous (1994) EPA report thrusts dioxins back into the spot light. ENDS ref. no. 236, pp 21–24

Anonymous (1997) Industry news: cluster rule signed. Tappi J 80(12):14

Archibald F (2000) The presence of coliform bacteria in Canadian pulp and paper mill water systems—a cause for concern? Water Qual Res J Can 35(1):1–22

Axegard P, Berry RM, Gellerstedt G, Lindblad PO, Luthie CE, Popke I, Voss RH, Wrist PE (1993) The effects of recent changes in bleached softwood Kraft mill technology on organochlorine emissions: an international perspective. In: Jameel H (ed) Bleaching, vol 2. Tappi Press, Atlanta Georgia, pp 759–770

Bajpai P, Bajpai PK (1996) Organochlorine compounds in bleach plant effluents—genesis and control. PIRA International, Leatherhead, Surrey, UK

Bajpai P (2012) Biotechnology for pulp and paper processing. Springer Science+Business Media, New York, p 414

Baruah BK (1997) Effect of paper mill effluent on plankton population of wetland. Environ Ecol 15(4):770–777

Canada Gazette (1992) Part II, vol 126(11)

Dutta SK (1999) Study of the physicochemical properties of effluent of the paper mill that affected the paddy plants. J Environ Pollut 6(2 and 3):181–188

Erisction G, Larsson A (2000) DNA A dots in perch (Perca fluviatillis) in coastal water pollution with bleaching in pulp mill effluents. Ecotoxicol Environ Saf 46:167–173

Felder DP, D'surney SJ, Rodgers JH, Deardorff TL (1998) A comprehensive environmental assessment of a receiving aquatic system near an unbleached Kraft mill. Ecotoxicology 7:313–324

Folke J, Renberg L and McCubbin N (1996) Environmental aspects of ECF vs. TCF Pulp Bleaching. In: Servos MR, Munkittrick KR, Carey JH, Van Der Kraak GJ (eds) Environmental fate and effects of pulp and paper mill effluents. St. Lucia Press, Delray Beach, Florida, p 681

Gavrilescu D (2006) Environmental consequences of pulp and paper manufacture 1. Bleached Kraft pulp mills. Environ Eng Manage J 5(1):37–49

Gupta A (1997) Pollution load of paper mill effluent and its impact on biological environment. J Ecotoxicol Environ Monit 7(2):101–12

Haglind I, Hultman B, Stromberg L (1993) SSVL environment 93—results from a research project conducted by the Swedish Pulp and Paper Industry. In: Proceedings TAPPI Environmental Conference, Boston, Massachusetts, Book 1, p 245

Hewitt LM, Carey JH, Dixon DG, Munkittrick KR (1996) Examination of bleached kraft mill effluent fractions for potential inducers of mixed function oxygenase activity in rainbow trout. In: Servos MR, Munkittrick KR, Carey JH, Van Der Kraak GJ (eds) Environmental fate and effects of pulp and paper mill effluents. St. Lucie Press, Delray Beach, p 79

Higachi R, Cherr G, Shenker J, Macdonald J, Crosby D (1992) A polar high molecular mass constituent of bleached Kraft mill effluent is toxic to marine organisms. Environ Sci Technol 26(12):2413–2420

Hodson PV, Servos MR, Munkittrick KR, Carey JH, Van Der Kraak (1996) Mixed function oxygenase induction by pulp mill effluents: advances since 1991. In: Servos MR, Munkittrick KR, Carey JH, Van Der Kraak GJ (eds) Environmental fate and effects of pulp and paper mill effluents. St. Lucie Press, Delray Beach, p 349

Howe J, Michael RW (1998) Effects of pulp mill effluent irrigation on the distribution of elements in the profile of an arid region soil. Environ Pollut 105:129–135

Johansson NG, Clark FM and Fletcher DE (1995) New technology developments for the closed cycle bleach plant. In: Proceedings international non-chlorine bleaching conference, Amelia Island Florida, 5–9 Mar 1995

King HM, Baldwin DS, Rees GN, Mcdonald S (1999) Apparent bioaccumulation of Mn derived from paper mill effluent by the freshwater crayfish cherax destructor—the role of Mn oxidising bacteria. Sci Total Environ 226:261–267

Kovacs TG, Martel PH, Voss RH (2002) Assessing the biological status of fish in a river receiving pulp and paper mill effluents. Environ Pollut 2002(118):123–140

Mandal TN, Bandana TN (1996) Studies on physicochemical and biological characteristics of pulp and paper mill effluents and its impact on human beings. J Freshw Biol 8(4):191–196

Martel PH, Kovacs TG, O'Connor BI, Voss RH (1994) Water Res 28:1835

McCubbin N (1989) Dioxin '89: a bit like the adventures in alice and wonderland. Pulp Pap Can 90(11):17–19

McCubbin N, Sprague JB, Bonsor N (1990) Kraft mill effluents in Ontario. Pulp Pap Can 91(3):T112–T114

McDonough TJ (1995) Recent advances in bleached chemical pulp manufacturing technology Part 1. Tappi J 78(3):55–62

Muller EF, Halliburton D (1990) Characterization of effluent quality at seven Canadian bleach
    kraft pulp mills. Chemosphere 20(7–9):743
NCASI (1995) Characterization of effluent quality at seven Canadian bleach kraft pulp mills
    operating complete substitution bleach plants. In: National council of the paper industry for
    air and stream improvement, May 1995
Nelson PJ (1998) Elemental chlorine free (ECF) and totally chlorine free (TCF) bleaching of
    pulps. In: Young, RA, Akhtar M (ed) Environmentally friendly technologies for the pulp and
    paper industry. Wiley, New York, p 215
O'connor BI, Kovacs TG, Voss RH, Martel PH, van Lierop B (1994) A critical review of scandi-
    navian studies on the aquatic impacts of pulp and paper effluents. Pulp Pap Can 95(3):99–108
O'connor B, Kovacs T, Gibbons S, Strang AL (2000) Carbon dioxide in pulp and paper mill
    effluents from oxygen -activated sludge treatment plants as a potential source of distress and
    toxicity to fish. Water Qual Res J Can 35(2):189–200
Owens JW (1991) A critical review of scandinavian studies on the aquatic impacts of pulp and paper
    effluents. In: Proceedings TAPPI Environmental Conference, Book 1, San Antonio, Texas, p 271
Pokhrel D, Viraraghavan T (2004) Treatment of pulp and paper mill wastewater—a review. Sci
    Total Environ 333(1–3):37–58
Rappe C, Wagman N (1995) Trace analysis of PCDDs and PCDFs in unbleached and bleached
    pulp samples. Organohalogen Compd 23:377–382
Rennel J (1995) TCF—an example of the growing importance of environmental perceptions in
    the choice of fibers. Nord Pulp Pap Res J 10(1):24–32
Savant DV, Abdul-Rahman R and Ranade DR (2006) Anaerobic degradation of adsorb-
    able organic halides (AOX) from pulp and paper industry wastewater. Bioresour Technol
    97:1092–1104
Schnell A, Hodson PV, Steel P, Melcer H and Carey JH (1993) In: Proceedings CPPA environ-
    mental conference
Singh RS, Marwaha SS, Khanna PK (1996) Characteristics of pulp and paper mill effluents. J Ind
    Pollut Control 12(2):163–172
Skipperud L, Salbu B, Hagebo E (1998) Speciation of trace elements in discharges from the pulp
    industry. Sci Total Environ 217:251–256
Sodergren A (1989) In: Biological effects of bleached pulp mill effluent. National Swedish
    Environmental Protection Board Report, p 3558
Sodergren A, Bengtsson BE, Jonsson P, Lagergren S, Larson A, Olsson M, Renberg L (1988)
    Water Sci Technol 20(1):49
Sodergren A, Adolfsson-Erici M, Bengtsson BE, Jonsson P, Lagergren S, Rahm R and Wulff F
    (1993) Environmental impact of bleached pulp mill effluent. In: Sodergren, A (ed) Bleached
    pulp mill effluent composition, fate and effects in Baltic Sea. Environmental protection
    agency report 4047. Arlow, p 26–46
Solomon K, Bergman H, Huggett R, Mckague B, Mackay D (1996) A review and assessment
    of the ecological risks associated with the use of chlorine dioxide for the bleaching of pulp.
    Pulp Pap Can 97(10):345–354
Stepanova L, Lindstrom-Seppa P, Hanninen OOP, Kotelevtsev SV, Glaser VM, Novikow CN
    (2000) Lake Baikal: biomonitoring of pulp and paper mill wastewater. Aquat Ecosyst Health
    Manage 3:259–269
Sundelin B (1988) Effects of sulphate pulp mill effluents on soft bottom organisms—a micro-
    cosm study. Water Sci Technol 20(2):175–177
Suntio LR, Shiu WY, Mackay DA (1988) A review of the nature and properties of chemicals pre-
    sent in pulp mill effluents. Chemosphere 17(7):1249–1299
Vice K, Carrol R (1998) The cluster rule: a summary of phase I. Tappi J 81(2):91–98
Wayland M, Trudeau S, Marchant T, Parker D, Hobson KA (1998) The effect of pulp and paper
    mill effluent on an insectivorous bird, the tree swallow. Ecotoxicology 7:237–251
Yen NT, Oanh NTK, Reutergard LB, Wise DL, Lan LTT (1996) An integrated waste survey and
    environmental effects of COGIDO, a bleached pulp and paper mill in Vietnam on the receiv-
    ing water body. Global Environ Biotechnol 66:349–364

Chapter 5
Strategies Used for Reducing the Generation of Pollutants

Some excerpts taken from Bajpai (2012a). Biotechnology for Pulp and Paper Processing with kind permission from Springer Science+Business Media.

The pulp and paper industry has taken great strides in recognizing and solving many environmental problems by adopting the following:

1. Process modification through adaptation of cleaner technologies as alternatives to conventional technologies.
2. End-of-pipe pollution treatment technologies, which are essential either as a supplement or as a backup measure to pollution reduction techniques in order to meet the effluent regulation standards.

Both these two approaches are equally important in meeting environmental regulations and are presented in separate sections.

5.1 Pollution Reduction Through Process Modification

5.1.1 Nonconventional Pulping Technologies

Several alternate pulping technologies have been developed but these technologies have not yet reached the point where they are fully commercially viable. These technologies have multiple advantages when compared to other popular methods such as Kraft and sulfite pulping. In particular, the ability to obtain relatively high quality lignin adds value to a process stream otherwise considered as waste.

5.1.1.1 Organosolv Pulping

Organosolv is a pulping technique that uses an organic solvent to solubilize lignin and hemicelluloses (Young and Akhtar 1998). It has been considered in the context of both pulp and paper manufacture and biorefining for subsequent conversion of cellulose to fuel ethanol. The process was invented by Theodore Kleinert as

P. Bajpai, *Bleach Plant Effluents from the Pulp and Paper Industry*, SpringerBriefs in Applied Sciences and Technology, DOI: 10.1007/978-3-319-00545-4_5, © The Author(s) 2013

an environmentally benign alternative to Kraft pulping. Organosolv has several advantages when compared to other popular methods such as Kraft or sulfite pulping. In particular, the ability to obtain relatively high quality lignin adds value to a process stream otherwise considered as waste. Organosolv solvents are easily recovered by distillation leading to less water pollution and elimination of the odor usually associated with Kraft pulping. Organic solvents such as methanol, ethanol, and other alcohols are used for pulping. This process is economical for small- to medium-scale mills with significant recovery of chemicals for reuse. However, pulping must be conducted in enclosed containers to prevent the loss of volatile solvents and for workers' safety. Major benefits include the elimination of odorous sulfur containing compounds in the effluents and air.

The Organosolv pulping processes can be classified as organic acid based, alcohol/water based, and mixed processes which use inorganic and organic pulping chemicals. None of the proposed Organosolv processes has been implemented successfully on a mill scale. A full-scale Organocell plant in Kelheim, Germany, had to be shut down due to insolvable problems. The Repap Enterprises Alcell mill at Newcastle, NB, Canada had a similar fate. It is too early to make a conclusive judgment about the alternatives but it is clear that Organosolv pulping is not accepted by modern technology. Information on the environmental impact of the Organosolv pulping processes is scarce and is mainly derived from pilot-scale trials. Whether the advantages can be achieved in full-scale plants is to a certain extent doubtful or at least not proven.

5.1.1.2 Organic Acid Pulping

Organic acid processes are alternative methods of organosolv pulping to delignify lignocellulosic materials to produce pulp for paper (Poppius et al. 1991). Typical organic acids used in the acid pulping methods are formic acid and acetic acid. The process is based on acidic delignification to remove lignin. The pulping operation can be carried out at atmospheric pressure. Acid used in pulping can be easily recovered by distillation and re-used in the process. Cellulose, hemicellulose and lignin can be effectively separated by degradation in aqueous acetic acid or formic acid. The cooking liquor is washed from the pulp, and both cooking chemicals and water are recovered and recycled completely. Formic acid can also be used to enhance acetic acid pulping. The temperature and pressure can be lower when formic acid is used in pulping compared to those used in alcohol or acetic acid pulping. Organic acid lignin is an optimal feedstock for many value-added products, due to its lower molecular weight and higher reactivity. The Organosolv pulping processes based on organic acid cooking are the Milox, Acetosolv, and Formacell processes. The Milox process is an Organosolv pulping process which uses peroxyformic acid or peroxyacetic acid as the cooking chemical (Leponiemi 2008). Peroxyformic or peroxyacetic acids are simple to prepare by equilibrium reaction between hydrogen peroxide and formic or acetic acids. These are highly selective chemicals that do not react with cellulose or other wood polysaccharides in the same way as formic acid. The hydrogen peroxide consumption is reduced by performing the process in two or three stages. The two-stage formic acid/peroxyformic acid process can be used to produce high viscosity and fully bleached (90 % ISO) pulp with a reasonable yield (40–48 %). The pulping stages are carried out at atmospheric pressure and at temperatures below 100 °C. The resulting pulps have kappa numbers between 5 and 35 (Muurinen 2000). The hydrogen peroxide charge needed can be reduced by using a three-stage cooking method.

Acetic acid was one of the first organic acids used for the delignification of lignocellulosic raw material to produce pulp for paper. Processes based on the use of acetic acid as an organic solvent have been applied with success to hard and softwoods, and even to non-wood materials. It can be used as a pulping solvent in uncatalyzed systems (Acetocell method) or in catalyzed systems (the Acetosolv method).

5.1.1.3 Biopulping

The concept of biopulping is based on the ability of some white-rot fungi to colonize and degrade selectively lignin in wood thereby leaving cellulose relatively intact. There are certain process conditions and design requirements necessary to gain a biopulping effect (Akhtar et al. 1998). Biopulping can be carried out in bioreactors of different types, including open chip piles, depending on the requirements of the particular microorganism would have for optimal results. High moisture content (around 55–60 %) should be kept in wood chips during the biotreatment step to ensure an optimal colonization and penetration of fungal hyphae. The degree of asepsis should be controlled to ensure a successful wood colonization by the particular fungal strain used depending on its resistance against contamination and ability to compete with the microbial biota existing in the wood chips. This process appears to have the potential to overcome some problems associated with conventional chemical pulping methods. Biochemical pulping reduces the amount of cooking chemicals, increases the cooking capacity, or enables extended cooking, resulting in lower consumption of bleaching chemicals. Increased delignification efficiency results in an indirect energy saving for pulping, and reduces pollution. The waste load produced by biopulping should be considerably lower and more benign than effluents currently produced by conventional pulping methods. Fungal pretreatment also reduces the pitch content in the wood chips and improves the pulp quality in terms of brightness, strength, and bleachability. The bleached biopulps are easier to refine than the reference pulp. The process has been scaled up toward industrial level, with optimization of various process steps and evaluation of economic feasibility. The process can be carried out in chip piles or in silos. Biopulping technology has advanced rapidly within recent years and pilot mill trials have been started worldwide (Reid et al. 2010). This technology coincides perfectly with environmentally safe production strategies and can be implemented in existing production plants without major changes.

5.1.2 Internal Process Modifications

In order to minimize the formation of chlorinated organic material, and to ensure compliance with various regulations on AOX and dioxin limits, the producers of bleached Kraft pulp have made substantial changes in the pulp manufacturing technology (McDonough 1995). As the quantity of chlorine is a function of lignin content of the pulp, a lower lignin content before chlorination stage will contribute toward reduction in chlorine requirements. Several methods have been tried to reduce the lignin content of the pulp. These include:

- Extended delignification and oxygen delignification
- Improved pulp washing
- Oxidative alkali extraction stage
- Substitution of elemental chlorine with chlorine dioxide
- Ozone ECF, totally chlorine-free bleaching, and chloride removal processes
- Bleach filtrate recovery
- Enzyme pre-treatment, and
- Fungal pre-treatment.

Extended Delignification and Oxygen Delignification

Extended delignification and oxygen delignification remove more lignin from the wood before the unbleached pulp enters the bleach plant (Bajpai 2012b). Therefore, fewer bleaching chemicals are required, less organic waste is generated in the bleaching process, less waste treatment is necessary, and discharges per ton of pulp manufactured are lower. Energy use also is lower because additional organic material removed from the pulp can be burned in the recovery boiler instead of being discharged, and because more heated process water is recirculated within the mill. To extend delignification in the pulping process, new digesters can be installed or existing digesters can be modified to increase the length of time that wood chips are cooked. This removes more lignin without compromising the strength of the pulp. The addition of certain chemicals such as anthraquinone in the pulping stage can have a similar effect (McDonough 1995). Extended cooking removes 35–40 % more lignin than conventional cooking. The limiting factors, for extending the delignification, are the pulp viscosity and its quality.

Oxygen delignification systems employ oxygen to remove additional lignin after the wood chips have been cooked in the digester but before the pulp enters the bleach plant. The filtrates from the pulp washers following the oxygen delignification step are routed to the chemical recovery system. It is important to note that mills worldwide currently using TCF or ozone-ECF bleaching technologies, also employ extended delignification, oxygen delignification, or both. The one chloride removal technology now being tested in a mill-scale demonstration is designed for mills with an ECF process that also uses oxygen delignification. The removal of additional lignin prior to the bleaching process is an essential foundation for the cost-effective operation of these technologies. Without the removal of additional lignin using extended delignification or oxygen delignification prior to bleaching, too much material is present for the cost-effective use of the oxygen-based bleaching compounds or chloride removal processes. Oxygen bleaching was the first step taken to achieve the zero-discharge bleach plant in Sweden. Today, all the Swedish Kraft pulp mills have installed the oxygen bleaching process. In North America, most mills have installed oxygen delignification system. Being cheap, non-toxic, renewable, and widely available reagent, oxygen is an excellent alternative to conventional polluting inorganic chemicals.

Improved Pulp Washing

The pulp mill section of paper industries normally uses brown stock washers for extraction of black liquor and for washing of pulp. The washing efficiency of these washers depend on the nature and quality of the fibrous raw materials. Most small industries use brown stock washers for washing of pulp produced from agro residues but the efficiency of these washers is not satisfactory as high carry over of black liquor along with pulp is observed in agro-based industries. Since the pulp from agro residues is difficult to dewater, the pulp and paper industry can use the modified washing systems such as belt filter press, double wire washer etc., to minimize the amount of black liquor entering the bleaching section.

5.1.2.3   Oxidative Alkali Extraction

Addition of oxygen to the pulp in alkaline extraction is an efficient method for increasing the bleaching effect and decreasing the consumption of chlorine chemicals and hence the pollution load. Oxygen improves the dissolution of lignin. An oxygen-reinforced extraction stage is designated EO. When peroxide is added, it is designated EOP. The extraction process reinforced with atmospheric peroxide has gained favor due to its good effectiveness and low capital requirement for implementation. However, peroxide reinforcement can be more effective when pressurized in the so-called PHT -stage technology (Pereira et al. 1995; Breed et al. 1995). The most common types of alkaline extractions nowadays are those reinforced with oxygen and peroxide, partially pressurized (EOP) or pressurized all the way, the so-called (PO)-stage. Use of hydrogen peroxide in the extraction stage results in more reduction in chlorine chemicals and also decreases the color of mill effluent (Anderson 1996). For a softwood Kraft pulp (kappa factor 0.18), the use of oxygen and hydrogen peroxide in the extraction stage (EOP) results in reduction in kappa number after extraction from 3.5 to 2.5. Other advantages of alkaline extraction with oxidants include improvement of environmental parameters such as color COD, BOD and AOX (Smook 1992). Significant reduction in effluent color is the greatest benefit of H_2O_2 addition in the extraction stage. It is common to use the alkaline extraction reinforced with oxygen (EO) or hydrogen peroxide (EP, PHT) or both (EOP, PO), to compensate for lower chlorine dioxide availability and also to make possible the bleaching in short sequences. Bleach plants that have a low availability of chlorine dioxide and require high peroxide dosages (0.8–1.0 %) need more severe conditions for peroxide consumption. In these cases, pressurized peroxide stages such as (PO) or PHT are recommended, because they allow use of high temperatures. A very significant increase in brightness is achieved when peroxide is applied to the alkaline extraction for eucalyptus Kraft pulp. Another positive effect is the peroxide effectiveness to bleach shives; even when they are not completely bleached, they are lighter and less visible (Anderson 1996). It is also possible to use H_2O_2 in the second alkaline extraction stage to counteract pulp darkening to reduce chlorine dioxide consumption.

A more powerful oxygen extraction stage is accomplished by raising the temperature in the stage, increasing the oxygen charge, pressurizing the pre-retention tube, and adding hydrogen peroxide. The most important factor is temperature. Use of pressurized peroxide stages (PO) makes it possible to achieve a high final brightness in totally chlorine free (TCF) bleaching (Bajpai 2012b). In sequences with chlorine dioxide (ECF bleaching), a powerful peroxide stage will reduce the consumption of chlorine dioxide or even replace one chlorine dioxide stage. A hot, pressurized peroxide stage operates at temperatures above 100 °C with a small amount of oxygen added. A prerequisite for successful peroxide bleaching is that the content of metal ions, e.g., manganese, copper, and iron, is low. Several mills around the world use oxygen-peroxide reinforced extraction stage.

Substitution of Chlorine Dioxide for Elemental Chlorine

Some bleached Kraft pulp mills are improving the quality of their effluent by replacing elemental chlorine with chlorine dioxide. The substitution of chlorine dioxide for 100 % of the elemental chlorine used in the bleaching process is one form of ECF bleaching. This improved bleaching process reduces the formation of many chlorinated organic compounds during the bleaching process. However, the quantity of effluent from the mill is not reduced. Further progress in reducing the quantity and improving the quality of the effluent ultimately depends on installing an improved pulping process or one of the technologies described below. Other technologies that reduce effluent quantity may become available in the future. Mills also operate ECF bleaching processes with improved pulping processes, such as oxygen delignification and/or extended delignification. The substitution of ClO_2 for Cl_2 was shown to reduce the formation of AOX by 80 %, which may be a reasonable way to satisfy the existing AOX regulation. Further reduction in AOX discharge by addition of dimethylsulfoxide along with ClO_2 substitution was reported (Lachenal et al. 1996). Parthasarathy et al. (1994) showed that even with 70 % substitution at the first stage, 2,3,7,8-TCDD and 2,3,7,8-TCDF concentrations in the effluent were not detectable. Similar conclusion is applicable to 100 % ClO_2 substitution at the first stage. Switching to 100 % ClO_2 substitution would result in AOX reduction of as much as 75 %. The AOX, from the bleach plant, was reduced to less than 1.8 kg/ton of pulp. There was also a substantial decrease of chloroform in the untreated effluent. After the treatment, the chloroform concentration was found to be below the detection limit.

Ozone Based ECF, TCF, and Chloride Removal Processes

The main effect of using chlorine and/or chlorine dioxide in the bleaching process is that chlorides in the bleach plant filtrates make the filtrates to be sent to the chemical recovery system corrosive. So wastewater from the bleach plant that contains chlorinated compounds is not sent through the chemical recovery system, but is treated and discharged into receiving waters. Replacing chlorine compounds in the bleaching process with oxygen-based compounds reduces the corrosiveness of the wastewater from each stage of the bleaching process in which the substitution is made. This allows bleach plant filtrates to be sent back through the mill's chemical recovery system and reused instead of being treated and discharged. One way to remove chlorides is to substitute ozone for chlorine or chlorine dioxide in the first stage of the bleaching sequence, thus allowing the filtrates from the first bleaching and extraction stages to be recirculated to the recovery boiler.

In the last stage of ozone-based ECF bleaching systems, chlorine dioxide is used to brighten the pulp (Bajpai 2012b). This is a low-effluent process because only the last bleaching stage uses fresh water that is discharged to the treatment plant; the ozone stage removes most of the remaining lignin. TCF bleaching processes go one step further than ozone-ECF processes to replace all chlorine compounds in the bleaching process with oxygen-based chemicals such as ozone or hydrogen peroxide. TCF processes currently offer the best opportunity to recirculate the filtrates from the entire bleach plant because they have eliminated chlorine compounds from all bleaching stages; however, few mills currently operate TCF processes in a low-effluent mode.

Commercial-scale TCF processes are relatively new. Mills installing these processes typically discharge the filtrates when they first install the processes, and plan to move to low-effluent processes over time. Add-on technologies that remove the chlorides from the mills' process water using additional evaporating and chloride removal equipment are in earlier stages of development. Rather than substitute bleaching compounds like ozone for chlorine dioxide, these processes do not reduce the use of chlorine dioxide, but seek to remove chlorides from wastewater with additional processing steps. Unlike the ozone ECF or TCF processes, the chloride removal processes generate an additional waste product that must be disposed.

TCF bleaching has been studied extensively over the last few years. The use of oxygen-based chemicals in lieu of chlorine containing bleaching agents not only decreases the amount of chlorinated organic material in bleaching effluent, but also results in effluent which is almost free from corrosive components.

Installing pollution-prevention technologies at bleached Kraft pulp mills reduces releases to the environment and the environmental impacts from the mill's effluent. Because hardwoods have lower lignin contents, the estimates of AOX and COD for hardwood bleach plant filtrates with traditional ECF bleaching will be similar to those of softwood bleach plant filtrates with enhanced ECF. In traditional ECF bleaching processes, all of the remaining lignin in the unbleached pulp is removed in the bleaching process and leaves the mill in the effluent. Mills that employ enhanced ECF and low-effluent technologies recirculate more filtrates that contain wood waste to the chemical recovery system, and less organic waste leaves the mill in the effluent. With enhanced ECF processes, for example, about 50 % of the remaining lignin is removed during the oxygen delignification or extended delignification step.

Bleach Filtrate Recovery

Apart from the internal measures tried, Mapple et al. (1994a, b) discussed bleach filtrate recovery, directed toward bleach plant closure (zero-discharge). Laboratory studies indicated that by using bleach plant closure effluent volume, color, AOX, and BOD could be reduced by 50, 90, 85, and 70 %, respectively. Evans et al. (1994) showed that, for the existing mills, it would be economically beneficial to eliminate the use of gaseous chlorine and hypochlorite to minimize the chlorine input, and to evaporate and incinerate the bleach plant waste separately. Though the implementation of recovery of the bleach plant effluent is considered technically possible today, there are a number of areas where additional developmental work has to be carried out to reduce the risks that are involved in implementing these new concepts. These areas include product quality development, management of non-process elements, reduction of solid waste generation and air emissions, management of process upset conditions, and mill chemical balances.

Enzyme Pretreatment

Xylanase pretreatment of pulps prior to bleach plant reduces bleach chemical requirements and permits higher brightness to be reached (Viikari et al. 2009). The reduction in chemical needed can translate into significant cost savings when high levels of chlorine dioxide and hydrogen peroxide are being used. A reduction in the use of chlorine chemicals clearly reduces the formation and release of chlorinated organic compounds in the effluents and the pulps themselves. The ability of xylanases to activate pulps and increase the effectiveness of the bleaching chemicals may allow new bleaching technologies to become more effective. This means that xylanase pretreatment may eventually permit expensive chlorine-free alternatives such as ozone and hydrogen peroxide to become cost effective. Traditional bleaching technologies also stand to benefit from xylanase treatments. Xylanases are easily applied and require essentially no capital expenditure. Because chlorine dioxide charges can be reduced, xylanase may help eliminate the need for increased chlorine dioxide generation capacity. Similarly, the installation of expensive oxygen delignification facilities may be avoided. The benefit of a xylanase bleach boosting stage can also be taken to shift the degree of substitution toward higher chlorine dioxide levels while maintaining the total dosage of active chlorine. Use of high chlorine dioxide substitution dramatically reduces the formation of AOX. In TCF-bleaching sequences, the addition of enzymes increases the final brightness value, which is a key parameter in marketing chlorine-free pulp. In addition, savings in TCF bleaching are important with respect both to costs and to the strength properties of the pulp. The production of TCF pulp has increased dramatically during recent years. Several alternative new bleaching techniques based on various chemicals such as oxygen, ozone, peroxide and peroxyacids have been developed. In addition, an oxygen delignification stage has already been installed at many Kraft mills. In the bleaching sequences in which only oxygen-based chemicals are used, xylanase pretreatment is generally applied after oxygen delignification to improve the otherwise lower brightness of the pulp or to decrease bleaching costs. The TCF sequences usually also contain a chelating step in which the amount of interfering metal ions in pulp is decreased. It has been observed that the order of metal removal (Q) and enzymatic (X) stages is important for an optimal result. When aiming at the maximal benefit of enzymatic treatment in pulp bleaching, the enzyme stage must be carried out prior to or simultaneously with the chelating stage. In fact, the neutral pH of enzyme treatment is optimum in many cases for chelation of magnesium, iron, and manganese ions that must be removed before bleaching with hydrogen peroxide. The TCF technologies applied today are usually based on bleaching of oxygen-delignified pulps with enzymes and hydrogen peroxide.

Oxidative enzymes from white-rot fungi can directly attack lignin. These enzymes are highly specific toward lignin; there is no damage or loss of cellulose and their use can produce larger chemical savings than xylanase, but this method has yet not been developed to full scale (Bajpai 2006). As compared to oxygen delignification, treatment with lignin-oxidizing enzymes results in more removal of lignin. This translates into substantial savings of energy and bleaching chemicals which in turn leads to a lower pollution load. Lignin-oxidizing enzymes are not currently available in sufficient quantity for mill trials, and scale-up of enzyme production from fungal cultures is costly. Cloning of genes for lignin-oxidizing enzymes has been reported and may provide an alternative production route. Experience with xylanase and with other enzymes has shown that enzymes can be successfully introduced in the plant. Thus, oxidative enzymes which can be regarded as catalysts for oxygen and hydrogen peroxide driven delignification, may also find a place in the bleach plant in coming years.

Fungal Pretreatment

Pre-treatment with fungi has been shown to replace up to 72 % of the chemicals needed to bleach Kraft pulp (Fujita et al. 1991). Only a few white-rot fungi have been tested for their ability to delignify Kraft pulps. In Japan, a 5-day fungal (F) treatment of hardwood Kraft pulp with strain IZU-154 replaced the CE_1DE_2D sequence with the FCED sequence, resulting in a 72 % chlorine saving. Nishida et al. (1995) investigated the bio-bleaching of hardwood unbleached Kraft pulp by *Phanerochaete chrysosporium* and *Trametes versicolor*, and established a positive correlation between the decrease in kappa number and increase in the brightness of the fungal treated pulp. Very few researchers have measured the impact of fungal bleaching on the effluent quality. Fujita et al. (1991) reported 50 and 80 % reductions in COD and color loading, respectively, in FCED bleaching sequence. Despite the emphasis on fungal bleaching, as a means to reduce the use of chlorine and associated formation of chlorinated organics, the effect upon chlorinated organics has not been reported.

5.2 External Treatment

Process modifications and allied solutions can reduce the pollution load, but not to the extent that waste generation is totally eliminated. The wastes generated will still require treatment, in order to meet the prescribed effluent standards, before disposal into the environment. The technologies that can be used to treat pulp and paper mill effluent (end-of-pipe remedies) include physico-chemical processes, electrochemical processes, enzymatic treatment, and biological treatment processes.

5.2.1 Physico-Chemical Processes

A variety of physico-chemical methods have been tried for the treatment of pulp and paper mill effluent. These include coagulation, flocculation, settling, adsorption on active surfaces such as fly ash, and membrane techniques.

A wide variety of coagulants have been studied for their effectiveness in the removal of color. Coagulation, with alum dosage of 100 mg/L, has been reported to reduce 80 % color and 50 % COD (Dilek and Goekcay 1994). Through the use of a mixture of polyethylene and modified starches, Milstein et al. (1991) reported 75, 59, and 80 % removals of AOX, COD, and color, respectively. Chloride and sulfate salts of iron and aluminum were effective in treating bleach Kraft mill effluents (Stephenson and Duff 1996). Removal efficiencies of 88 and 98 % were observed for total organic carbon (TOC) and turbidity, respectively. Toxicity was also markedly reduced. With chitosin as coagulant, 90 and 70 % reduction of color and TOC, respectively, were reported (Ganjidoust et al. 1996). Tong et al. (1999) and Ganjidoust et al. (1997) compared horseradish peroxide (chitosan) and other coagulants such as aluminum sulfate, hexamethylene diamine epichlorohydrin polycondensate, polyethyleneimine, to remove AOX, TOC, and color. The authors found that modified chitosan was very effective in removing these pollutants than other coagulants. Wagner and Nicell (2001) investigated the treatment of foul condensate, defined by phenolic compounds, and toxicity using microtox assay from Kraft pulping by horseradish peroxidase and H_2O_2 and found a total phenol reduction below 1 mg/L and toxicity (microtox assay) reduction by 46 %. Rohella et al. (2001) reported that polyelectrolytes were better than the conventional coagulant alum to remove turbidity, COD, and color. Indian researchers (Sheela and Distidar 1989) reported on black liquor treatment by precipitation with $CaSO_4.2H_2O$ in the presence of CO_2. The removal of dissolved solids was reported to be 63 %. On the other hand, Wang and Pan (1999) reported that the use of coagulants such as polyethylene oxide worsened the settleability and increased COD levels, turbidity, and suspended solids of the treated effluent when the dose was between 25 and 250 ppm. Coagulation with aluminum sulfate or modified adsorbents was the best option for color removal from the sulfate and sulfite wood pulp and paper industry, according to Chernoberezhskii et al. (1994).

Adsorption on active surfaces has frequently been employed for removing pollutants, such as biphenyls, organochlorines, and heterocyclic organics. Fly ash as an adsorbing medium has been reported to remove COD and color efficiently (Nancy et al. 1996). Murthy et al. (1991) observed a high removal of color by the use of activated charcoal, fuller's earth, and coal ash. Sullivan (1986) reported that the wastewater produced by the Union Camp Facility at Franklin, VA, can be treated by activated carbon and ion exchange to reduce color and chloride to levels acceptable for reuse. Another research group (Shawwa et al. 2001) reported 90 % removal of color, COD, DOC, and AOX from bleached wastewater by the adsorption process, using activated coke as an adsorbent. Das and Patnaik (2000) investigated the lignin removal efficiency of the blast furnace dust (BFD) and slag by the adsorption mechanism. It was found that 80.4 and 61 % lignin were removed by BFD and slag, respectively. Narbaitz et al. (1997) reported that the PACT™ process was effective in removing AOX from the Kraft mill effluent to meet AOX regulations.

Ultrafiltration has been reported to be a good method for removing colored material from bleach Kraft mill effluent. With this the technique, Pejot and Pelayo (1993) have achieved 79–91 % decolorization and 74–88 % COD removal and Yao et al. (1994) achieved 90 and 99 % reduction in TOC and AOX, respectively. Brite (1994) studied nanofiltration, combined with electrodialysis, at pilot-scale level to treat pulp bleach effluent. He reported over 95 % removal of the contaminating toxic organic halides, salts, and colorants, and the treated effluents were found suitable for process reuse. Jonsson et al. (1996) studied the treatment of paper coating color effluent treatment by membrane filtration (MF). They suggested that the composition of the color had a significant influence on the performance. Membrane separation techniques were reported to be suitable for removing AOX, COD, and color from pulp and paper mills (Afonso and Pinho 1991; Falth 2000). De Pinho et al. (2000) made a comparison of ultrafiltration and ultrafiltration plus dissolved air flotation. They found 54, 88, and 100 % removal of TOC, color, and SS, respectively, by ultrafiltration alone and 65, 90, and 100 % removal of TOC, color, and SS, respectively, by Ultrafiltration plus dissolved air flotation. Merrill et al. (2001)

found that MF, and granular membrane filtration (GMF) removed heavy metals from the pulp and paper mill wastewaters. Dube et al. (2000) observed 88 and 89 % removal of BOD, and COD, respectively, by reverse osmosis (RO).

Oxidation can accomplish the destruction of both chromophoric and toxic compounds. Oxidants that have been used or proposed include chlorine (Clark et al. 1994), oxygen (Sun et al. 1992), ozone (Hostachy et al. 1996), and peroxide (Smith and Frailey 1990). Ozonation was reported to remove 72 % of the effluent color at a dosage of 40 ppm. Further, it was found to selectively destroy acute toxicity of chemomechanical pulp effluents (Roy-Arcand and Archibald 1991b). Hostachy et al. (1996) reported complete detoxification of BKME at low ozone doses (0.5–1.0 kg/ton air dried pulp). Korhonen et al. (2000) observed 90 % removal of EDTA and a 65 % removal of COD by ozone treatment of the pulp mill effluent. Oeller et al. (1997) found high removal of COD and DOC from the pulp effluent by ozone treatment. Yeber et al. (1999) reported considerable removal of COD, TOC, and toxicity from pulp mill effluent and increased biodegradability of the effluent after treatment with ozone. Freire et al. (2000) reported a 12 % reduction of TOC, 70 % reduction of total phenols, and 35 % reduction of colors from bleached pulp mill effluent after 60 min of ozonation.

Sun et al. (1992) removed approximately 70–80 % of TOCl and 60–70 % of the effluent color associated with high molecular weight chlorolignins by oxidation at high temperatures under alkaline conditions. Clark et al. (1994) reported 50–90 % decolorization of the effluent with chlorine. The cost of the oxidizing agent is a significant issue. The known inexpensive oxidants (e.g., chlorine and hypo) are also the ones that produce unwanted organochloride byproducts, especially chloroform. The others are, typically, either very expensive or unstable or both. To make these processes more economical, it has been proposed to use them as a pretreatment to biological treatment. Pretreatment partially degrades the compounds, that otherwise resist biological treatment, into forms that are biodegradable, thus eliminating color and toxicity. Studies have found that pretreatment with ozone or peroxide does increase the biodegradability of Kraft mill caustic extraction stage effluent.

Some investigators have found beneficial effects from using ozone or peroxide in combination with ultraviolet light in treating bleachery waste. Color removal efficiencies of as high as 80 % were achieved in a pilot-scale treatment of bleach Kraft mill effluent through oxidation, first with peroxide (using a dose as low as 480 mg/L), and then with UV radiation (Smith and Frailey 1990). These researchers employed UV radiation, ozone, and peroxide to treat bleach mill effluent and reduced the color to 1.5 kg/ton (from an initial value of 3.5 kg/ton).

Physico-chemical technologies are costly and rather unreliable. Oxidation using ozone and hydrogen peroxide may prove costly. The coagulation/precipitation methods of treatment produce voluminous sludges, handling of which poses difficulty. Oxidation using chlorine species (chlorine and hypo) are reported to generate secondary pollutants such as chloroform. MF techniques require pretreatment and are capital intensive. Membrane fouling is the another problem associated with these techniques.

5.2.2 Electrochemical Processes

In electrochemical treatment, chloride in the effluent is converted by electrolysis to chlorate, hypochlorite, and chlorine. The chlorine and hypochlorite oxidize the organic compound in the effluent and chloride gets regenerated. Springer et al. (1994) used a bench-scale electrochemical cell, in a study for investigating technical and economic feasibility of electrochemical treatment as a method for the removal of color and toxicity. Operating costs were observed to be between $0.50 and $2.32/1,000 gallons of effluent or $5–$23/ton of pulp.

Electrochemical systems are effective but high in operating costs because much of electrochemical energy is consumed in undesirable side reactions.

5.2.3 Advanced Oxidation Treatment Technologies

Nowadays, advanced oxidation processes (AOPs) are being studied for oxidizing the recalcitrant materials in wastewaters by means of the high oxidative power of the OH• radical (Hirvonen et al. 1996) associated with AOPs. Advanced oxidation processes are defined as near ambient temperature and pressure water treatment processes which are based on the generation of hydroxyl radicals (OH•) to initiate oxidative destruction of organics. The hydroxyl radical is a powerful, nonselective chemical oxidant which reacts typically a million to a billion times faster than ozone and hydrogen peroxide resulting in greatly reduced treatment costs and system size. AOPs are characterized by production of OH• radicals and selectivity of attack which is a useful attribute for an oxidant. The versatility of AOP is also enhanced by the fact that they offer different possible ways for producing OH• radicals. Hydroxyl radicals can be created in reactions involving Ozone; Hydrogen peroxide; Ozone + hydrogen peroxide; Photooxidation; Photocatalysis; Electron beam irradiation; or Sonolysis.

AOPs offer the potential for a complete destruction of hazardous organic compounds in process wastewater without generating secondary pollution. AOPs are best suited for destroying toxic organic solutes in solutions with low suspended solids and low concentration of organic contaminants. AOPs can be used as stand-alone treatment processes, or as post- or pre-treatment steps to conventional processes as part of an integrated treatment system.

Photocatalytic treatment of bleaching effluent from the pulp mill has been studied by Toor et al. (2007) in a low-cost, non-concentrating shallow pond slurry reactor using artificial UV light with Degussa P25 TiO_2 as catalyst. The results obtained with sunlight for the UV radiation were similar to those obtained under artificial UV light of the same intensity. The degradation of more than 90 % of the pollutants in terms of COD was possible. This research has shown that efficient degradation of pulp and paper mill effluent containing recalcitrant compounds is possible by this method. Experiments under solar radiation yielded results comparable to those obtained indoors under UV lamps which show that this process can be applied on an industrial scale and its cost may not be prohibitive. Since pulp and paper mills already use holding ponds for microbiological treatment of wastewaters, large-scale shallow pond reactors for solar detoxification may be an option on the front or back end of a combined solar/microbiological treatment system for wastewater.

Kamwilaisak and Wright (2012) investigated laccase enzyme and titanium dioxide for lignin degradation. Laccase from *T. versicolor* served as the bio-catalyst, and TiO_2 served as the photocatalyst. The catalysts were used in single- and dual-step configurations. For comparison, lignin degradation by laccase and Titania alone were studied. Operational conditions were 50 ± 1 °C, pH 5.0, and with a lignin concentration (molecular weight of 16,000–175,000) of 1.0 g/L. H_2O_2 was used as a mediator to increase laccase and TiO_2 degradation ability. The results show that H_2O_2 plays a significant role in improving lignin degradation by TiO_2 and that 100 % decolorization and delignification was achieved. Gas chromatography–mass spectrometry analysis confirmed the presence of organic acids as a prominent compound class in TiO_2/H_2O_2 processes. It was found that not only can laccase and TiO_2 completely degrade lignin but also the process yields highly desirable byproducts, such as succinic and malonic acids.

Xu et al. (2007) treated the bleaching effluent from a wastepaper pulp mill with the solar photo-Fenton process in a laboratory-scale reactor. The treatment involved a constant intensity of irradiation with a solar simulator of 250 W xenon lamps at different pH and temperatures as well as initial concentration of hydrogen peroxide and iron. TOC removal of over 90 % was observed at optimum conditions, particularly at high temperature. The large-scale application of the process has strong potential for the efficient removal of organics from the effluent. This can be realized with the use of economical irradiation source of light and the high temperature of the original effluent.

The photocatalytic treatment of wastewaters from board paper industries with the photo-Fenton reagent or semiconductor TiO_2 resulted in a considerable decrease of the organic pollutants load (Amat et al. 2005). This could be achieved using solar light and has been scaled up to a pilot plant. The best performance was obtained for wastewaters from higher degree of water circuit closure, when COD values were quite high.

Heterogeneous photocatalysis and ozone treatment have been successfully used to remove low concentrations of organic and halo-organic contaminants in bleached Kraft mill effluents (Torrades et al. 2001). A 20-h photocatalytic treatment using TiO_2 (Degussa P 25) -predominantly anatase removed all of the color and most of the TOC, AOX, and COD in a lightly loaded effluent obtained from the D-stage of an AOD bleach sequence. For an effluent with a higher contaminant load obtained from the first D-stage in a conventional ECF sequence, the best treatment is the one that takes advantage of a previous treatment with lime and the synergistic effect of simultaneous photocatalysis and ozonation. Similar levels of TOC removal were obtained in the sequences hydrolysis–Ph–O_3 and hydrolysis–O_3–Ph, although for larger reaction times. The TOC, COD, and AOX reductions in the treated effluents were above 80 % in all cases, and the effluents were fully decolorized.

The present treatment costs of photocatalytic systems are slightly higher than those of conventional techniques, but the efforts being made in the design of more efficient systems with improved catalyst usage will establish this technology to be a cleaner and cost-effective alternative.

5.2.4 Treatment with Enzymes

Enzyme-based treatment offers some distinct advantages over physical and chemical decolorization and AOX precipitation methods. These advantages are that only catalytic and not stoichiometric amounts of the reagents are needed, and the low organic concentrations and large volumes typical of bleaching effluents are therefore less of a problem. Also, both complete microbial systems and isolated enzymes have been shown to reduce the acute toxicity by polymerizing and thereby rendering less soluble many of the low molecular mass nonchlorinated and polychlorinated phenolics (Bollag et al. 1988; Klibanev and Morris 1981; Ruggiero et al. 1989).

Hakulinen (1988) published a review on the use of enzymes for waste water treatment in the pulp and paper industry. The new possibilities of using enzymes like laccase, peroxidase and ligninase for this effect were examined. Forss et al. (1987) examined the use of laccase for effluent treatment. They aerated pulp bleaching waste water in the presence of laccase for 1 h at pH 4.8 and subsequently flocculated with aluminum sulfate. High removal efficiencies were obtained for chlorinated phenols, guaiacols, vanilins and catechols. Roy-Arcand and Archibald (1991a) studied direct dechlorination of chlorophenolic compounds in pulp and paper mill effluent by laccases from *T. versicolor* and found that all the major laccases, secreted by *T. versicolor*, could partially dechlorinate a variety of chlorophenolics. These researchers also studied effects of horseradish peroxidase (HRP) and *P. chrysosporium* peroxidase on the mixture of five chlorophenolics (pentachlorophenol, tetrachloroguaicol, 4,5,6-trichloroguaiacol, 4,5-dichloroguaiacol, 2,4,6-trichlorophenol). Both peroxidase enzymes were found to degrade the majority of substrates except pentachlorophenol. But the *P. chrysosporium* peroxidase was superior to both HRP and laccase in degrading pentachlorophenol and inferior to HRP in degrading the other four phenolics.

Paice and Jurasek (1984) studied the ability of HRP to catalyze color removal from bleach plant effluents. The color removal from effluents at neutral pH by low levels of hydrogen peroxide was enhanced by the addition of peroxidase. No precipitation occurred during the decolorization process. The catalysis with peroxidase (20 mg/L) was observed over a wide range of peroxide concentrations (0.1–800 mM) but the largest effect was between 1 and 100 mM. The pH optimum for catalysis was around 5.0. Compared with mycelial color removal by *Coriolus versicolor*, the rate of color removal by peroxide plus peroxidase was initially faster (for the first 4 h) but the extent of color removal after 45 h was higher with the fungal treatment. Further addition of peroxidase to the enzyme-treated effluents did not produce additional catalysis. Thus, the peroxide/peroxidase system did not fully represent the metabolic route used by the fungus. One working hypothesis has been proposed to explain the behavior of enzymes in the decolorization process (Paice and Jurasek 1984). Glucose is used by the cell to produce peroxidase which is one of the extracellular enzymes often found in white-rot fungi. This enzyme oxidizes the chromophores and so removes the color from bleaching waste water.

Field (1986) patented a method for the biological treatment of waste waters containing non-degradable phenolic compounds and degradable non-phenolic compounds. It consisted of an oxidative treatment to reduce or eliminate toxicity of the phenolic compounds followed by an anaerobic purification. This oxidative pretreatment could be performed with laccase enzymes and it was claimed to reduce COD by 1000 fold. Call (1991) patented a process on the use of laccase for wastewater treatment. He claimed that wastewater from delignification and bleaching could be treated with laccases in the presence of nonaromatic oxidants and reductants and aromatic compounds. Almost complete polymerization of the lignins is obtained which is 20–50 % above the values attainable with the addition of laccase alone. About 70–90 % lignin is developed into insoluble form, which is removed by flocculation and filtration.

Milstein et al. (1988) described the removal of chlorophenols and chlorolignins from bleaching effluents by combined chemical and biological treatments. The organic matter from spent bleaching effluents of the chlorination, extraction, or a mixture of both stages, was precipitated as a water insoluble complex with polyethyleneimide. The color, COD, and AOX were reduced by 92, 65, and 84 %, respectively, for the chlorination effluent and by 76, 70, and 73 % for the extraction effluent. No significant reduction in BOD of treated effluent was detected but fish toxicity was greatly reduced. Enzyme treatment results in coprecipitation of the bulk mono-and dichlorophenols with the liquors of the chlorination and extraction bleaching stages. Lyr (1963) reported that laccase of *T. versicolor* partially dechlorinates PCP and Hammel and Tordone (1988) reported that peroxidase from *P. chrysosporium* can partially dechlorinate PCP and 2,4,6-trichlorophenol.

Though the use of enzyme-based treatments offers some distinct advantages over physical and chemical methods in that only catalytic amounts of reagents are needed, biochemical instability and difficulty in reusing the enzyme are its disadvantages. Immobilization of the enzymes is required for biochemical stability and reuse of the enzymes. Carbon immobilized laccase was used by Davis and Burns (1992) to decolorize extraction stage effluent at the rate of 115 PCU/enzyme unit/h. The removal rate was found to increase with the increasing effluent concentration. Dezotti et al. (1995) developed a simple immobilization method where activated silica was used as a support and used it for enzymatic color removal from extraction stage effluent by lignin peroxidase (LiP) from *Chrysonilia sitophila* and by commercial HRP. Immobilized HRP gave 73 % decolorization and LiP gave 65 and 12 % reductions in COD and color, respectively. Immobilized enzymes were found to retain activity even after 5 days of contact with the Kraft mill effluent. Ferrer et al. (1991) reported that immobilized lignin peroxidase decolorized Kraft effluent. It has been claimed that novel lignin peroxidases produced by *P. chrysosporium* mutant strain SC 26 decolorize bleaching effluents (Farrell 1987a, b).

Karimi et al. (2010) investigated the efficiency of AOPs, enzymatic treatment, and combined enzymatic/AOP sequences for the color remediation of soda and chemimechanical pulp and paper mill effluent. The results indicated that under all circumstances, AOP using ultraviolet irradiation (photo-Fenton) was more efficient in the degradation of effluent components than the dark reaction. It was found that both versatile peroxidase (VP) from Bjerkandera adusta and laccase from *T. versicolor*, as pure enzymes, decolorize the deep brown effluent to a clear light-yellow solution. In addition, it was found that in the laccase treatment, the decolorization rates of both effluents were enhanced in the presence of 2, 2'-azinobis (3-ethylbenzthiazoline-6-sulfonate), while in the case of VP, Mn(+2) decreased the efficiency of the decolorization treatment. The concomitant use of enzymes and AOPs has considerable effect on the color remediation of effluent samples.

5.2.5 Treatment with Bacteria

Bacterial treatments include aerobic treatment, anaerobic treatment and combination of both treatments. Combinations of anaerobic and aerobic treatment processes are found to be efficient in the removal of soluble biodegradable organic pollutants (Pokhrel and Viraraghavan 2004; Begum et al. 2012).

Aerobic Treatment

The most common aerobic biological methods used in the treatment of pulp mill effluents are:

Aerated lagoon treatment (ASB)

The aerated lagoon is a low rate aerobic biological process and is the oldest and simplest type of aerobic biological treatment system to construct and operate. Extensive experience in applying ASBs in the treatment of pulp mill effluent is available. In both Canada and the US, most of the early constructed secondary treatment systems in pulp and paper mills, where available land space is not limited, are aerated lagoons (Wilson and Holloran 1992; Turk 1988). In developing countries, lagoons are the major process for the treatment of pulp mill effluent. Removals of AOX from bleached Kraft mill effluents are quite variable among systems, ranging from 15 to 60 % with an average of 30 % (Wilson and Holloran 1992).

Significant work has been done by Yin (1989) and Bryant et al. (1987, 1988) to determine the mechanism of AOX removal in aerated lagoons. AOX removal occurs by biosorption of organohalides to biomass and anaerobic dehalogenation and degradation in the benthyl layer, of the lagoon with biosorption providing the transport mechanism. Both high and low molecular weight chlorolignins are reported to adsorb to aerobic biomass but aerobic dehalogenation has not been reported. Conversely, it has been suggested that the majority of AOX removal in an aerated lagoon is due to aeration enhanced hydrolytic splitting of chlorine from the organic substrate (Yin 1989). MLSS levels in an aerated lagoon are too small to allow significant biosorption to sludge. Removal of resin and fatty acids (RFAs) in CTMP effluent is generally >95 % (Liu et al. 1996). Analysis of relative removals of different MW fractions in three North American ASBs was reported (Bryant 1990). Low molecular weight AOX was removed more effectively (43–63 %) than high molecular weight AOX (4–31 %). Effluent AOX removal from mills using hardwood and softwood was comparable but furnish changeovers reduced the removal performance. In a separate lab-scale ASB study, degradation of hardwood derived TOCl was greater (44–52 %) than for softwood derived TOCl (44 %) (Yin et al. 1989).

ASBs have been widely employed in the treatment of Kraft mill effluent, TMP and CTMP effluents (Tomar and Allen 1991; McCubbin 1983; Liu et al. 1996; Johnson and Chatterjee 1995; Saunamaki et al. 1991; Jokela et al. 1993) for the removal of BOD and toxicity, chlorophenols, low molecular weight AOX, RFAs.

Experiments on recirculation of biomass in aerated lagoons have indicated that a four-fold increase in lagoon biomass could increase removal efficiency from 50 to 60 % (Boman et al. 1988).

Bryant et al. (1997) observed 67 % removal of ammonia from black liquor spill at temperatures of 22–35 °C and pH near 7.3 in an aerated lagoon. Junna and Ruonala (1991) reported removal of BOD7 ranging between 50 and 75 % and chlorinated phenolics between 10 and 50 % in an aerated lagoon. Welander et al. (1997) reported COD removal of 30–40 % in a fullscale lagoon and 60–70 % in a pilot-scale plant. Stuthridge and Mcfarlane (1994) found that 70 % removal of the AOX from an aerated lagoon was due to a short residence time section in that part of the treatment system where the chlorinated stage effluents were mixed with general mill wastewaters. The effect of simple mixing was reported to be responsible for 15–46 % removal. Chernysh et al. (1992) found large variations in AOX and TOC removal in a controlled batch study of bleached Kraft effluent in an operating lagoon under both aerobic and anaerobic conditions. Stuthridge et al. (1991) reported 65 % removal of AOX from bleached Kraft pulp and paper mill effluent.

Achoka (2002) reported that an oxidation pond removed chemical compounds greater than 50 %. Schnell et al. (2000) found reduction in BOD, AOX, chlorinated, and polychlorinated phenolics from an aerated lagoon.

Fulthrope and Allen (1995) studied the ability of three bacterial species to reduce AOX in bleached Kraft mill effluents. *Ancylobacter aquaticus* A7 exhibited the broadest substrate range but could only affect significant AOX reduction in softwood effluents. *Methylobacterium* CP13 exhibited a limited range but was capable of removing significant amounts of AOX from both hardwood and softwood effluents. By contrast, *Pseudomonas sp.* Pl exhibited a limited substrate range and poor to negligible reductions in AOX levels from both effluent types. Mixed inocula of all the three species combined and inocula of sludge from mill treatment systems removed as much AOX from softwood effluents as did pure populations of *Methylobacterium* CP13. Rogers et al. (1975) treated the bleached Kraft mill effluent in a bench scale aerated lagoon for 29, 58, and 99 h, and showed that toxicity, BOD and resin acids were most consistently reduced during the 99 h treatment. Leach et al. (1978) reported the biodegradation of seven compounds representing the major categories of toxicants in a laboratory-scale batch aerated lagoon. Resin acids (major source of acute toxicity) were readily biodegradable but only part (less than 30 %) of the load of chlorophenolic compounds was removed. Deardorff et al. (1994) reported that the efficiency of AOX removal through biotreatment of combined bleach plant effluent increases with increasing chlorine dioxide substitution. Biological treatment in an aerated lagoon reduced the concentration of polychlorinated phenolic compounds by 97 %. Jokela et al. (1993) reported that aerobic lagoon systems removed 58–60 % of the AOX from the water phase whereas the full-scale activated sludge plants removed 19–55 %. Eriksson and Kolar (1985) have shown that high molecular weight fraction compounds in bleach plant effluents cannot be degraded in an aerated lagoon. In another study, it was shown that chloroform is stripped during the biological treatment and COD, AOX and high molecular weight material are reduced to a lesser extent (SSVL-85 Project 4, Final report).

Reduction of individual chlorinated organics from aerated basins has been reported (Boman et al. 1988; Saunamaki et al. 1991; Lindstrom and Mohamed 1988; Wilson and Holloran 1992; Brynt et al. 1987; Gergov et al. 1988; Voss 1983). Individual removal efficiencies for various chlorophenols range from 30 to 89 %. Information obtained from Paprican has indicated removal efficiency up to 100 % for chlorovanillins (Willson and Holloran 1992).

ASBs provide distinct advantages over high rate systems such as ASTs, including little or no nutrient addition required, expect at initial start up, lower net settleable solids generation, lower energy consumptions due to avoidance of sludge handling and reduced aeration requirement, and better toxicity removal.

Activated sludge treatment

AST is a high rate biological process and has been used by the pulp and paper industry when the available land space is small or a low treated effluent suspended solids concentration is required. It is reported that ASTs generally remove much higher quantities of AOX than aerated lagoons. Removal efficiencies of 14–65 % have been reported. A number of full-scale AST systems are operated in the United States and in Canada for the treatment of various pulp mill effluents, including those from Kraft, paper board, deinking, TMP and CTMP, sulfite and newsprint mill operations (Buckley 1992; Paice 1995; Johnson and Chatterjee 1995).

When the bleaching effluents from chlorination and extraction stage were treated in an activated sludge process, the AOX reduction was found to be 30–40 % in 8 days. About 70–80 % of the total AOX reduction was achieved in about 4 days (Mortha et al. 1991). The presence of high molecular weight material in the bleached Kraft effluent was found to improve the removal of chlorophenolic compounds. Growth experiments using microorganisms from a lab-scale activated sludge reactor showed that high molecular weight material had a significant role in soluble COD and chlorophenol removal (Bullock et al. 1994). Large decreases in the soluble COD and increases in the biomass were observed with the addition of high molecular weight materials to the low molecular weight fraction. The addition of mono- and dichlorinated phenolic compounds at concentrations up to 10 mg/L were found to have no effect on the metabolism or growth of the microorganisms in the activated sludge. While 6-chlorovanillin, 2,4-dichlorophenol, and 4,5 dichloroguaiacol were found to be stable in uninoculated controls and in inoculated low molecular weight effluent over a 160 h period, these compounds decreased significantly, when low molecular weight with three times the original concentration of high molecular weight material was inoculated with microorganisms. Gergov et al. (1988) investigated pollutant removal efficiencies in mill-scale biological treatment systems. They found that 48–65 % of AOX was removed by the activated sludge process.

Pilot-scale investigation of activated sludge treatment of bleached Kraft effluent at a Northern Ontario mill site was studied by Melcer et al. (1995). The AS system was operated at 1 day HRT, 25–30 days SRT, and 30 °C. Treated effluents were found to pass all the toxicity tests. A high level of effluent quality was achieved with low concentrations of AOX (4–13 mg/L), total chlorophenolics (0.3–0.32 mg/L), toxicity equivalents (0.4–5 mg/L), total RFAs (0–4 mg/L), BOD (4–12 mg/L), and soluble COD (142–274 mg/L) being recorded over the whole period of investigation.

Mohamed et al. (1989) reported removal of chlorinated phenols, 1,1-dichlorodimethyl sulfone (DDS), and chlorinated acetic acids in an oxygen activated sludge effluent treatment plant. Knudsen et al. (1994) reported a high reduction of BOD and soluble COD by a two-stage activated sludge process. Chandra (2001) observed efficient removal of color, BOD, COD, phenolics, and sulfide by microorganisms—*Pseudomonas putida*, *Citrobacter sp.*, and *Enterobacter sp.* in the activated sludge process. Shere and Daly (1982) claimed that TMP wastewater was readily degradable by the activated sludge process. Kennedy et al. (2000) reported that activated sludge was successful in removing nearly all detectable Microtox™ toxicity from bleached Kraft pulp mills. Junna and Ruonala (1991) reported 90 % BOD7, 70 % COD, 40–60 % AOX, and 60–95 % chlorinated phenols removal by the activated sludge process. Hansen et al. (1999) suggested upgrading the activated sludge plant by the addition of Floobeds in series because it was found that they increased COD and BOD removal from 51 to 90 and 70 to 93 %, respectively. Raghuveer and Sastry (1991) reported that a minimum of mixed liquor suspended solids of 2,000–2,500 mg/L and an aeration time of 6–8 h were required to remove 83–88 % of BOD. High removals of BOD, COD, AOX, and chlorinated phenolics have been achieved in the activated sludge process. Bryant et al. (1992) found AOX removal of 46 % on average from two activated sludge systems studied. Andreasan et al. (1999) suggested the addition of an anoxic selector before the activated sludge plant to improve the sludge settleability problem.

Valenzuela et al. (1997) studied the degradation of chlorophenols by *Alcaligenes eutrophus* TMP 134 in bleached Kraft mill effluent. After 6 days of incubation, 2,4-dichlorophenoxyacetate (400 ppm) or 2,4,6 trichlorophenol (40–100 ppm) were extensively degraded. In short-term incubations, indigenous microorganisms were unable to degrade such compounds. Degradation of 2,4,6-trichlorophenol by strain JMP 134 was significantly lower at 200–400 ppm of compound. This strain was also able to degrade 2,4-dichlorophenoxyacetate, 2,4,6-trichlorophenol, 4-chlorophenol and 2,4,5-trichlorophenol, when the effluent was amended with mixtures of these compounds. On the other hand, the chlorophenol concentration and the indigenous microorganisms inhibited the growth and survival of the strain in short-term incubations. In long-term incubations, strain JMP 134 was unable to maintain a large, stable population, but an extensive 2,4,6-trichlorophenol degradation was still observed. When combined effluents of a Kraft pulp mill were treated in a lab-scale activated sludge system, the average TOC and AOX removal efficiencies were found to be 83 and 21 %, respectively (Ataberk and Gokcay 1997).

The combined effects of oxygen delignification, ClO_2 substitution, and biological treatment on pollutants in bleach plant effluents were examined. Biological treatment did not reduce color but reduced COD, BOD, AOX, and toxicity (Graves et al. 1993). ClO_2 substitution reduced the discharge of all five pollutants with a large reduction in AOX. Oxygen delignification reduced discharges of the five pollutants, and effluents were easier to treat by aerobic methods. Treatment of bleaching effluent in sequential activated sludge and nitrification systems revealed that dechlorination of bleaching effluent took place in both systems (Altnbas 1997). In the activated sludge system, released inorganic chloride was 4.5–7 mg/L at a TOC loading rate of 0.03–0.07 mg/mg VSS/d; but it was decreased from 10 to 3 mg/L at a TOC loading rate of 0.006–0.06 mg/mg VSS/d. Removal efficiencies for individual chlorinated organics range from 18 to 100 % and are presented in Table 5.1.

Table 5.1 Activated sludge removal efficiencies for chlorophenols

Compound	Reduction (%)
Dichlorophenols	78
Trichlorophenols	51–69
Tetrachlorophenols	86–100
Pentachlorophenols	50–80
Dichloroguaiacols	67–97
Trichloroguaiacols	18–97
Tetrachloroguaiacols	59–99
Dichlorocatechols	37
Trichlorocatechols	63–95
Tetrachlorocatechols	59–90
Monochlorovanillins	94
Dichlorovanillins	100

Based on Wilson and Holloran (1992), Gergov et al. (1988), Saunamaki (1989), Rempel et al. (1990) and Mcleay (1987)

Liu et al. (1996) demonstrated that AOX removal mechanism includes biodegradation, adsorption to biomass, and air oxidation. Among these three, biodegradation is the major mechanism. Apart from achieving high AOX removals in ASTs, high performance COD, BOD, and TSS removals was recorded (Goronzy et al. 1996).

AOX removal efficiency was correlated to SRT and HRT (Rempel et al. 1990) in pilot-scale tests of air and oxygen activated sludge systems. The maximum reported AOX removal efficiencies (>40 %) were achieved for SRTs greater than 20 days and HRTs greater than 15 h. In a separate report on Finish activated sludge systems, the highest AOX removals (45 %) in mill scale units were reported for SRTs greater than 50 days (Salkinoja-Salonen 1990). Varying the HRTs and SRTs indicated that HRT had more of an effect on treatment performance than SRT. Longer HRTs led to improved BOD, COD, toxicity, and AOX removal, longer SRTs were not shown to significantly effect performance (Barr et al. 1996). Paice et al. (1996) investigated effluents from CMP/newsprint operation that was treated in two parallel laboratory-scale activated sludge systems. Removal of BOD and resin fatty acids in excess of 90 % was achieved with an HRT of 24 h. Anoxic conditioning of the sludge (Liu et al. 1997) and hydrolysis pretreatment of bleachery effluents (Zheng and Allen 1997) have been demonstrated to enhance AOX removal by about 8 and 20–30 %, respectively, in AST.

Tiku et al. (2010) studied the capability of three bacteria, *Pseudomonas aeruginosa* (DSMZ 03504), *P. aeruginosa* (DSMZ 03505), and *Bacillus megaterium* (MTCC 6544) to reduce the BOD and COD level of pulp and paper mill effluents within a retention time of 24 h in batch cultures. A concomitant reduction in TDS, AOX, and color (76 %) was also observed. This is the first report on the use of bacterial cultures for the holistic bioremediation of pulp mill effluent.

Rai et al. (2007) examined three lignin-degrading bacterial strains, identified as *Paenibacillus sp., Aneurinibacillus aneurinilyticus*, and *Bacillus sp.* for the treatment of pulp and paper mill effluent. The results of this study revealed that all three bacterial strains effectively reduced color (39–61 %), lignin (28–53 %), BOD (65–82 %), COD (52–78 %), and total phenol (64–77 %) within 6 days of incubation. However, the highest reduction in color (61 %), lignin (53 %), BOD (82 %), and COD (78 %) was recorded by *Bacillus sp.* while, maximum reduction in total phenol (77 %) was recorded with *Paenibacillus sp.* treatment. Significant reduction in color and lignin content by these bacterial strains was observed after 2 days of incubation, indicating that the bacterium initially utilized growth supportive substrates and subsequently chromophoric compounds thereby reducing lignin content and color in the effluent.

Mishra and Thakur (2010) isolated four different bacterial strains from pulp and paper mill sludge in which one alkalotolerant isolate having higher capability to remove color and lignin, was identified as *Bacillus sp.* by 16S RNA sequencing. Optimization of process parameters for decolorization was initially performed to select growth factors which were further substantiated by the Taguchi approach in which seven factors, % carbon, % black liquor, duration, pH, temperature, stirring, and inoculum size, at two levels, applying L-8 orthogonal array were taken.

Maximum color was removed at pH 8, temperature 35 °C, stirring 200 rpm, sucrose (2.5 %), 48 h, 5 % (w/v) inoculum size, and 10 % black liquor. After optimization, a 2-fold increase in color and lignin removal indicated the significance of the Taguchi approach.

Chandra et al. (2008) isolated eight aerobic bacterial strains from pulp paper mill waste and screened for tolerance of Kraft lignin (KL) using the nutrient enrichment technique in mineral salt media agar plate (15 g/L) amended with different concentrations of KL along with 1 % glucose and 0.5 % peptone (w/v) as additional carbon and nitrogen sources. The strains ITRC S6 and ITRC S8 were found to have the most potential for tolerance of the highest concentration of KL. These organisms were characterized by biochemical tests and further 16S rRNA gene sequencing, which showed 96.5 and 95 % sequence similarity of ITRC S(6) and ITRC S(8) and confirmed them as *Paenibacillus sp.* and *Bacillus sp.*, respectively. Among eight strains, ITRC S(6) and ITRC S(8) were found to degrade 500 mg/L of KL up to 47.97 and 65.58 %, respectively, within 144 h of incubation in the presence of 1 % glucose and 0.5 % (w/v) peptone as a supplementary source of carbon and nitrogen. In the absence of glucose and peptone, these bacteria were unable to utilize KL.

Monje et al. (2010) evaluated the aerobic and anaerobic biodegradability and toxicity to *Vibrium fischeri* of generated L-stage and total bleaching sequence effluents. The highest levels of aerobic and anaerobic degradation of the generated effluents were achieved for treatments with laccase plus violuric acid, with 80 % of aerobic degradation and 68 % of anaerobic biodegradation. *V. fischeri* toxicity was reduced for all the effluents after aerobic degradation.

Sequencing batch reactors

The SBR process is a fill and draw cyclic batch activated sludge process. Sequencing batch reactors (SBR) have the following advantages compared to conventional ASTs: lower operating costs as there is no aeration for 30–40 % of the total time, hence no sludge settler or recycling pumps are required. Control of filamentous bulking due to the anoxic fill, ability to tolerate peak flow and shock loads, and denitrification during the anoxic fill and settle stages. In addition, the control and operation of a SBR are flexible.

SBRs have been used for the treatment of pulp mill effluents and, in North America, there are several full-scale SBR systems treating various pulp mill effluents. SBRs generally produce smaller quantity of effluents than ASTs. Many authors have reported high removals of organic pollutants of Kraft mill wastewater by SBR treatment (Franta et al. 1994; Franta and Wilderer 1997; Milet and Duff 1998). Berube and Hall (2000) reported approximately 93 % removal of TOC by a membrane bioreactor. Asselin et al. (2000) found that suspended carrier biofilm reactor was effective in removing chronic toxicity from the effluent. Reid and Simon (2000) reported 100 % removal of methanol and 90 % removal of COD by SBR. Substantial removal of COD, TOC, BOD, lignin, and resin acids of TMP wastewater using high rate compact reactors at a retention time of 1.5 h have been reported (Magnus et al. 2000a, b). Magnus et al. (2000c) observed that a biological compact reactor gave 93 and 65 % removal of BOD and COD, respectively.

Removal of COD by a moving bed biofilm reactor had been demonstrated by Borch-Due et al. (1997). One of the major problems is the lack of experience for both design and operation of SBR systems for the treatment of such large quantities of effluents.

Other aerobic treatment systems

Other aerobic biological processes include rotary disc contractors and trickling filters. Mathys et al. (1993) and (1997) studied the treatment of CTMP mill wastewater in laboratory scale RBC. Application of these two processes for the treatment of pulp mill effluents are limited (Lunan et al. 1995; Mathys et al. 1997).

Anaerobic Treatment

The major anaerobic processes used for the treatment of pulp mill effluents include anaerobic lagoons, anaerobic contract processes, upflow anaerobic sludge blankets (UASB), anaerobic fluidized beds, and anaerobic filters. Anaerobic technologies are already in use for many types of forest industry effluents. UASB reactors and the contact process are the most widely applied anaerobic systems. Most of the existing anaerobic full-scale plants treat noninhibitory forest industry wastewater rich in readily biodegradable organic matter such as recycling waste water, and mechanical pulping (TMP) effluents. Full-scale application of anaerobic systems for chemical, semichemical, and chemithermomechanical, bleaching and debarking liquors is still limited.

The application of anaerobic treatment for degradation and dechlorination of Kraft bleach plant effluent has been studied by several researchers. The COD removals in the anaerobic treatment of bleaching effluents have ranged from 28 to 50 % (Lafond and Ferguson 1991; Raizer Neto et al. 1991; Rintala and Lepisto 1992). Removal of AOX was improved when easily degradable co-substrate was used to supplement the influent (Parker et al. 1993a). Many chlorophenolic compounds and chlorinated guaiacols were removed at greater than 95 % efficiency (Parker et al. 1993b). Fitzsimonas et al. (1990) investigated anaerobic dechlorination/degradation of AOX at different molecular masses in bleach plant effluents. A decrease in AOX was found with all molecular mass fractions. The rate and extent of dechlorination and degradation of soluble AOX decreased with increasing molecular mass. As high molecular weight chlorolignins are not amenable to anaerobic microorganisms, dechlorination of high molecular weight compounds may be due to combination of energy metabolism, growth, adsorption and hydrolysis.

Ali and Sreekrishnan (2007) treated black liquor and bleach effluent from an agroresidue-based mill anaerobically. Addition of 1 % w/v glucose yielded 80 % methane from black liquor with concomitant reduction of COD by 71 %, while bleach effluent generated 76 % methane and produced 73 and 66 % reductions in AOX and COD, respectively. In the absence of glucose, black liquor and bleach effluent produced only 33 and 27 % methane reduction, with COD reductions of 43 and 31 %, respectively.

Thermomechanical pulping of wastewater is found to be highly suitable for anaerobic wastewater treatment (Sierra-Alvarez et al. 1991; Jurgensen et al. 1985; Sierra-Alvarez et al. 1990). In a mesophilic anaerobic process, loading rates up to 12–31 kg $COD/m^3/d$ with about 60–70 % COD removal efficiency have been obtained (Sierra-Alvarez et al. 1990, 1991; Rintala and Vuoriranta 1988). In thermophilic anaerobic process conditions, up to 65–75 % COD removal was obtained at 55 °C at loading rate of 14–22 kg $COD/m^3/d$ in a UASB reactors (Rintala and Vuoriranta 1988, Rintala and Lepisto 1992).

Kortekaas (1998) studied anaerobic treatment of wastewaters from thermomechanical pulping of hemp. The wood and bark thermomechanical pulping waste waters were treated in a laboratory-scale UASB reactor. For both types of wastewaters, maximum COD removal of 72 % were obtained at loading rates of 13–16 g COD/l/d providing 59–63 % recovery of the influent COD as methane. The reactors continued to provide excellent COD removal efficiencies of 63–66 % up to a loading rate of 27 g COD/l/d, which was the highest loading rate tested. Batch toxicity assays revealed the absence of methanogenic inhibition by hemp TMP wastewaters, coinciding with the high acetolastic activity of the reactor sludge of approximately 1 g COD/g VSS/d.

Hall et al. (1986) and Wilson et al. (1987) studied anaerobic treatability of NSSC spent liquor together with other pulping and paper mill waste water streams. The methanogenic inhibition by NSSC spent liquor was apparently the effect of the tannins present in these wastewaters (Habets et al. 1985). Formation of H_2S in the anaerobic treatment of NSSC spent liquor has been reported but not related to methanogenic toxicity. Apparently, the evaporator condensates from the NSSC production are amenable to anaerobic treatment because of their high volatile fatty acid (Perttula 1991).

Unstable operations have been encountered in anaerobic treatment of pulp mill effluents. The reason for these problems are still unclear although it is believed that they may be associated with the toxicants in these effluents. Because of the unstable operation problems, application of anaerobic treatment technology in the paper industry sector is still limited. Research is underway to develop treatment systems that combine aerobic technology with the ultrafiltration process. The sequential treatment of bleached Kraft effluent in anaerobic fluidised bed and aerobic trickling filters was found to be effective in degrading chlorinated, high and low molecular material (Haggblom and Salkinoja-Salonen 1991). The treatment significantly reduced the COD, BOD and AOX of the waste water. COD and BOD reduction was greatest in the aerobic process whereas dechlorination was significant in the anaerobic process. With the combined aerobic and anaerobic treatment, over 65 % reduction of AOX and over 75 % reduction of chlorinated phenolics was observed. Measuring the COD/AOX ratio of the wastewater before and after treatment showed that the chlorinated material was as biodegradable as the non-chlorinated.

Dorica and Elliott (1994) studied the treatability of bleached Kraft effluent using anaerobic plus aerobic processes. BOD reduction in the anaerobic stage varied between 31 and 53 % with hardwood effluent. Similarly the AOX removal from the hardwood effluents was higher (65–71 %), for the single-stage and the two-stage treatment, respectively, than that for softwood effluents (34–40 %). Chlorate was removed easily from both softwood and hardwood effluents (99 and 96 %, respectively) with little difference in efficiency between the single-stage and two-stage anaerobic systems. At organic loadings between 0.4 and 1.0 kg $COD/m^3/d$, the biogas yields in the reactors were 0.16–0.37 L/g BOD in the feed. Biogas yield decreased with increasing BOD load for both softwood and hardwood effluents. Anaerobic plus aerobic treatment removed more than 92 % of BOD and chlorate. AOX removal was 72–78 % with hardwood effluents, and 35–43 % with softwood effluents. Most of the AOX was found to be removed from hardwood effluents during feed preparation and storage. Parallel control treatment tests in non-biological reactors confirmed the presence of chemical mechanisms during the treatment of hardwood effluent at 55 °C. The AOX removal that could be attributed to the anaerobic biomass ranged between 0 and 12 %. The Enso-Fenox process was capable of removing 64–94 % of the chlorophenol load, toxicity, mutagenicity and chloroform (Hakulinen 1982).

Haggblom and Salkinoja–Salonen (1991) found the sequential treatment of bleached Kraft effluent in an anaerobic fluidized bed and aerobic trickling filter effective in degrading chlorinated material. The treatment reduced the COD, BOD, and the AOX of the waste water. Reduction of COD and BOD was greatest in the aerobic process, whereas dechlorination was significant in the anaerobic process. With the combined aerobic and anaerobic treatment, over 65 % reduction of AOX and over 75 % reduction of chlorinated phenolic compounds was observed (Table 5.2). Microbes capable of mineralizing pentachlorophenol constituted approximately 3 % of the total heterotrophic microbial population in the aerobic trickling filter. Two aerobic polychlorophenol degrading *Rhodococcus* strains were able to degrade polychlorinated phenols, guaiacols and syringols in the bleaching effluent.

Singh (2007) and Singh and Thakur (2006) investigated sequential anaerobic and aerobic treatment in a two-step bioreactor for removal of color in the pulp

Table 5.2 Reduction of pollutants in anaerobic–aerobic treatment of bleaching effluent

Parameter	Reduction (%)
COD (mg O_2/l)	61
BioCOD (mg O_2/)l	78
AOX (mg Cl/l)	68
Chlorophenolic compound	
2,3,4,6 tetrachlorophenol	71
2,4,6 trichlorophenol	91
2,4 dichlorophenol	77
Tetrachloroguaiacols	84
3,4,5 trichloroguaiacols	78
4,5,6 trichloroguaiacols	78
4,5 dichloroguaiacols	76
Trichlorosyringol	64

Based on Haggblom and Salkinoja–Salonen (1991)

and paper mill effluent. In anaerobic treatment, color (70 %), lignin (25 %), COD (42 %), AOX (15 %), and phenol (39 %) were reduced in 15 days. The anaerobically treated effluent was separately applied in a bioreactor in presence of a fungal strain, *Paecilomyces sp.*, and a bacterial strain, *Microbrevis luteum*. Data has indicated reduction in color (95 %), AOX (67 %), lignin (86 %), COD (88 %), and phenol (63 %) by *Paecilomyces sp.* whereas *M. luteum* showed removal in color (76 %), lignin (69 %), COD (75 %), AOX (82 %), and phenol (93 %) by day third when 7 days anaerobically treated effluent was further treated by aerobic microorganisms.

Swedish MoDo Paper's Domsjo Sulfitfabrik is using anaerobic treatment at its sulfite pulp mill and produces all the energy required at the mill (Olofsson 1996). It also fulfills 90 % of the heating requirements of the inner town of Ornskoldvik. Two bioreactors at the mill transform effluent into biogas and slime. The anaerobic unit is used to 70 % capacity. A reduction of 99 % has been achieved for BOD_7 and the figure for COD is 80 %. There are plans to use the slime produced as a fertilizer.

In the Pudumjee Pulp and Paper Mill in India, the anaerobic pretreatment of black liquor reduced COD and BOD by 70 and 90 %, respectively, (Deshpande et al. 1991). The biogas produced is used as a fuel in boilers along with LSHS oil. The anaerobic pretreatment of black liquor has reduced organic loading at the aerobic treatment plant thereby reducing consumption of electrical energy and chemical nutrients.

A process based on UF and anaerobic and aerobic biological treatments has been reported (EK and Eriksson 1987; EK and Kolar 1989; Eriksson 1990). The UF was used to separate the high molecular weight mass, which is relatively resistant to biological degradation. Anaerobic microorganisms were believed to be able to more efficiently remove highly chlorinated substances than aerobic microorganisms. The remaining chlorine atoms were removed by aerobic microorganisms. The combined treatments typically removed 80 % of the AOX, COD, and chlorinated phenolics and completely removed chlorate (Table 5.3).

Anaerobic processes were previously regarded as being too sensitive to inhibitory compounds (Lettinga et al. 1990; Rinzema and Lettinga 1988). But now advances in the identification of inhibitory compounds and substances in paper

Table 5.3 Reduction of pollutants with ultrafiltration plus anaerobic/aerobic system and the aerated lagoon technique

Parameter	UF plus anaerobic/aerobic predicted reductions (%)	Aerated lagoon estimated reductions (%)
BOD	95	40–55
COD	70–85	15–30
AOX	70–85	20–30
Color	50	0
Toxicity	100	Variable
Chlorinated phenols	>90	0–30
Chlorate	>99	Variable

Based on Eriksson (1990), EK and Eriksson (1987), EK and Kolar (1989)

mill effluents as well as increasing insight into the biodegradative capacity and toxicity tolerance of anaerobic microorganisms has helped to demonstrate that anaerobic treatment of various inhibitory wastewaters is feasible. The capacity of anaerobic treatment to reduce organic load depends on the presence of considerable amounts of persistent organic matter and toxic substances. Most important toxicants are sulfate and sulfite (Pichon et al. 1988), wood resin compounds (Sierra-Alvarez and Lettinga 1990; McCarthy 1990), chlorinated phenolics (Sierra-Alvarez and Lettinga 1991), or tannins (Field and Lettinga 1991). These compounds are highly toxic to methanogenic bacteria at a very low concentration. In addition, a number of low molecular weight derivatives have also been identified as methanogenic inhibitors (Sierra-Alvarez and Lettinga 1991).

In CTMP wastewaters, resins and volatile terpenes may account for up to 10 % of the wastewater COD (1,000 mg/L) (Welander and Anderson 1985). The solids present in the CTMP effluent were found to contribute to 80–90 % of the acetoclastic inhibition (Richardson et al. 1991). The apparent inhibition by resin acids was overcome by diluting anaerobic reactor influent with water or aerobically treated CTMP effluent which contained less than 10 % of the resin acids present in the untreated wastewater (Habets and de Vegt 1991; MacLean et al. 1990). Similarly, inhibition by resin acids was overcome by diluting the anaerobic reactor influent with water and by aerating the wastewater to oxidize sulfite to sulfate prior to anaerobic treatment (Eeckhaut et al. 1986).

The AOX formed in the chlorination and alkaline extraction stages are generally considered responsible for a major portion of the methanogenic toxicity in bleaching effluents (Rintala et al. 1992; Yu and Welander 1988; Ferguson et al. 1990). Anaerobic technologies can be successfully applied for reducing the organic load in inhibitory waste waters if dilution of the influent concentration to subtoxic levels is feasible (Ferguson and Dalentoft 1991; Lafond and Ferguson 1991). Dilution will prevent methanogenic inhibition and favor microbial adaptation to the inhibitory compounds. Considerable dilution with other non-inhibitory waste streams such as Kraft condensates (Edeline et al. 1988) and sulfite evaporator condensates (Sarner 1988) prior to anaerobic treatment, has been shown to be effective for reducing this toxicity.

Tannic compounds present at fairly high concentrations, inhibit methanogenesis (Field et al. 1988, 1991). Dilution of wastewater or polymerization of toxic tannins to high molecular weight compounds by auto oxidation at high pH as the only treatment (Field et al. 1991) was shown to enable anaerobic treatment of debarking effluents.

5.2.6 Treatment with Fungi

Fundamental research on biological treatment of pulp mill wastewaters has been considered as an important field of study during the last three decades. The research indicates that white-rot fungi are the known microbes capable of degrading and decolorizing bleach plant effluents. White-rot fungi have been evaluated in trickling filters, fluidized bed reactors, and airlift reactors, at bench scale and found technologically feasible (Pellinen et al. 1988a, b; Prouty 1990). Only the mycelia color removal (MyCoR) process which uses *P. chrysosporium* to metabolise lignin color bodies, has crossed the bench scale and has been evaluated at the pilot-scale level (Campbell et al. 1982; Jaklin-Farcher et al. 1992) and found to be very efficient in destroying AOX. However, no reactors or processes studied so far have been found economically feasible because of the following reasons—energy required for lignins/chloro-lignin degradation by white-rot fungi has to be derived from an easily metabolizable, low molecular mass sugars; also, the process is not self sustaining from the angle of growth of white-rot fungi used.

Factors affecting fungal treatment of pulp mill effluents include concentration of nutrients and dissolved oxygen, pH, and temperature. Fungi, require certain essential minerals for their growth. Fungal decolorization involves a series of complex reactions many of which are catalyzed by enzymes. The addition of mineral solution activates the specific enzymes necessary for normal metabolism, growth, and decolorization. The fungus can tolerate a wide range of pH and temperature during decolorization compared to the growth stage. Decolorization is maximal under high oxygen concentration and the fungus requires a carbon source. A small addition of nitrogen is required to sustain decolorization because nitrogen is lost from the system by the extracellular enzyme secreted by the fungus (Bajpai 2012a).

To identify potential fungal strains for the treatment of bleach effluents, many researchers have screened cultures obtained from different sources. Japanese researchers (Fukuzumi et al. 1977) were probably the first to use white-rot fungi for effluent treatment. The fungi were grown in Erlenmeyer flasks in a liquid medium containing nutrients, vitamins, and spent liquor from the first alkali extraction stage of pulp bleaching. Among the fungi selected from 29 species of tropical fungi and 10 species of Japanese isolates, *Tinctoporia sp.* showed the highest decolorization. *Phlebia brevispora, Phlebia subserialis, Poria cinerascens*, and *T. versicolor* were tested by Eaton et al. 1982 and found to reduce the effluents color efficiently. In another study (Livernoche et al. 1983), 15 strains of white-rot fungi were screened for their ability to decolorize bleaching effluents. Five fungal strains—*T. versicolor, P. chrysosporium, Pleurotus ostreatus, Polyporus versicolor* and one unidentified strain showed decolorizing activity. Galeno and Agasin (1990) evaluated several white-rot fungi for their ability to decolorize bleaching effluents and found *Ramaira sp.* strain 158 to have the highest potential. Over 90 % of the color was removed after 140 h under air with a similar rate and extent of decolorization as *P. chrysosporium* did under oxygen.

The addition of an easily metabolizable nutrient is required to obtain the maximum decolorization efficiency with most white-rot fungi. However, this would increase the operational cost of the process. Moreover, if the added nutrients are not completely utilized during the decolorization stage, they could increase the BOD and COD of the effluents after fungal treatment. Esposito et al. (1991) and Lee et al. (1994) examined fungi that showed efficient decolorization of the extraction stage effluents without any addition of nutrients. Through a screening of 51 ligninolytic strains of fungi, the *Lentinus edodes* strain was shown to remove 73 % of the color in 5 days without any additional carbon source. Under these conditions, *L. edodes* was more efficient than the known *P. chrysosporium* strains (Esposito et al. 1991). Lee et al. (1994) screened fungi having high decolorization activity. The fungus KS-62 showed 70 and 80 % reduction of the color after 7 and 10 days of incubations, respectively. To obtain a reasonable basis for evaluation of an industrial fungal treatment, Lee et al. (1995) performed treatment of the extraction stage effluent with the immobilized mycelium of the fungus KS-62. This fungus showed 70 % color removal (initial color 6600 PCU) without any nutrient within 1 day of incubation with four times effluent replacement; however, the color removal started to decrease at the fifth replacement with the fresh extraction stage effluent. The decolorization activity of the fungus was restored by one replacement of extraction stage effluent containing 0.5 of glucose and high decolorization was continuously observed for four replacements in the absence of glucose. With the fungus KS-62, such decolorization activity was reportedly obtained for 29 days of total treatment period. As a result of screening 100 strains at low glucose concentration, *Rhizopus oryzae* (a *zygomycete*) and *Ceriporiopsis subvermispora* (a wood degrading white-rot fungi) were shown to remove 95 and 88 % of the color, respectively. Even in the absence of carbohydrates, significant color reductions were achieved (Nagarathnamma and Bajpai 1999; Nagarathnamma et al. 1999). Glucose has been found to be the most effective cosubstrate for decolorisation by most white-rot fungi (Nagarathanamma et al. 1999; Nagarathanamma and Bajpai 1999; Bajpai et al. 1993; Mehna et al. 1995; Fukuzumi 1980; Prasad and Joyce 1991; Bergbauer et al. 1991; Pallerla and Chambers 1995). Belsare and Prasad (1998) showed that decolorization by *Schizophyllum commune* could be rated in the following order: sucrose (60 %) glucose (58 %), cellulose (35 %), and pulp (20 %). With the fungus–*Tinctoporia*, ethanol was also found to be very effective cosubstrate for decolorization (Fukuzumi 1980). Ramaswamy (1987) observed that addition of 1 % bagasse pith as a supplementary carbon source resulted in 80 % color reduction in 7 days with *S. commune*. Eaton et al. (1982) compared the suitabilities of three primary sludges and combined sludge with that of cellulose powder for use as a carbon source for *P. chrysosporium* cultures. Archibald et al. (1990) reported that *T. versicolor* removed color efficiently in the presence of inexpensive sugar refining or brewery waste. With *R. oryzae* (Nagarathnamma and Bajpai 1999), maximum decolorization of the order of 92 % was obtained with addition of glucose in 24 h.

Eaton et al. (1980) studied the application of *P. chrysosporium* for the treatment of bleaching effluents. About 60 % decolorization of extraction stage effluent was found in shake flasks. The same mycelium could be recycled for up to 60 days or 6 successive batches. Mittar et al. (1992) also showed that under shaking conditions, the 7-day-old growth of the culture at 20 % (v/v) inoculum concentrations resulted in maximum decolorization (70 %) of the effluent along with more than 50 % reduction in BOD and COD.

Sundman et al. (1981) studied the reactions of the chromophoric material of extraction stage effluent during the fungal treatment without agitation. The results of these studies showed no preference toward degradation of lower molecular weight polymeric material over high molecular weight material. They noticed that the yield of high molecular weight material decreased to half during the fungal treatment. As the color also decreased by 80 %, they concluded that chromophores were destroyed. Further, they noticed that the fungal attack led to a decrease in the content of phenolic hydroxyl groups and to an increase in oxygen content.

Joyce and Pellinen (1990) proposed a process termed FPL NCSU MyCoR using *P. chrysosporium* for decolorization of pulp mill effluents. A fixed film MyCoR reactor is charged with growth nutrients which can include primary sludge as the carbon source and is inoculated with the fungus. The sludge will provide some of the required mineral nutrients and trace elements as well as carbon. Nitrogen rich secondary sludge can be also used to supply the nitrogen required for growth. After the mycelium has grown over the reactor surface, it depletes the available nitrogen and becomes ligninolytic. The reactor is then ready for use. Operations for over 60 days has been achieved in bench reactors in a batch mode. This process converts up to 70 % of the organic chlorides to inorganic chlorides in 48 h while decolorizing the effluent and reducing both COD and BOD by about half.

Huynh et al. (1985) used the MyCoR process for the treatment of chlorinated low molecular mass phenols of the extraction stage effluent. It was found that most of the chlorinated phenols and low molecular mass components of the effluent were removed during the fungal treatment. Pellinen et al. (1988b) have reported that the MyCoR process can considerably improve COD removal by simply using less glucose as the carbon source for the fungi—*P. chrysosporium*. However, the decolorization was reported to be faster at high glucose concentration. Yin et al. (1989) studied the kinetics of decolorization of extraction stage effluent with *P. chrysosporium* in an RBC under improved conditions. The kinetic model developed for 1 and 2 days retention times showed a characteristic pattern. The overall decolorization process can be divided into three stages viz. a rapid color reduction in the 1st hour of contact between the effluent and the fungus followed by a zero-order reaction and then a first-order reaction. The color removal rate on the second day of the 2 day batch treatment was less than that on the first day. The decolorization in a continuous flow reactor achieved approximately the same daily color removal rate, but the fungus had a longer working life than when in the batch reactor, thereby removing more color over the fungal life time. Pellinen et al. (1988a) studied decolorization of high molecular mass chlorolignin in first extraction stage effluent with white-rot fungus—*P. chrysosporium* immobilized on RBC. The AOX decreased almost by 50 % during one day treatment. Correlation studies suggested that decolorization and degradation of chlorolignin are metabolically connected.

The combined treatment of extraction stage effluent with white-rot fungi and bacteria have been also reported. Yin et al. (1990) studied a sequential biological treatment using *P. chrysosporium* and bacteria to reduce AOX, color, and COD in conventional softwood Kraft pulp bleaching effluent. In six variations of the white-rot fungus/bacterial systems studied, only the degree of fungal treatment was varied. In three of the six variations, ultrafiltration was also used to concentrate high molecular mass chlorolignins and to reduce effluent volume prior to fungal treatment. The best sequence, using ultrafiltration/white-rot fungus/bacteria, removed 71% TOCl, 50 % COD, and 65 % color in the effluent. Fungal treatment enhances the ability of bacteria to degrade and dechlorinate chlorinated organics in the effluent.

The degradation of model compounds—chlorophenols, and chloroguaiacols in pure water solution by fungal treatment using an RBC has been studied by Guo et al. (1990). It was found that at concentration of 30 mg/L, 80–85 % of chlorophenols and chloroguaiacols could be degraded after 3–4 h treatment.

Prouty (1990) proposed an aerated reactor in order to eliminate some of the problems associated with the RBC process. The fungal life in the aerated reactor was longer and the color removal rate was significantly higher than those of the RBC process in an air atmosphere. A preliminary economic evaluation of the RBC process indicated that the rate of decolorization and the life span of the fungus are the most critical factors (Joyce and Pellinen 1990). Yin et al. (1989) and Yin (1989) suggested that treatment of the extraction stage effluent by ultrafiltration before RBC treatment would be economically attractive. Their study also suggested that combining ultrafiltration and the MyCoR system could maximize the efficiency of the MyCoR process and reduce the treatment cost.

Although the MyCoR process was efficient in removing color and AOX from bleaching effluents, it also had certain limitations. The biggest problem was the short active life of the reactor. Therefore, several other bioreactors such as packed bed and fixed bed reactors were studied (Lankinen et al. 1991; Messner et al. 1990; Cammarota and Santanna 1992). The use of a trickling filter-type bioreactor with the fungus immobilized on porous carrier material was adopted in the MyCOPOR system (Messner et al. 1990). For extraction stage effluent with an initial color between 2600 and 3700 PCU, the mean rate of color reduction was 60 % during a 12 h run. The mean AOX reduction value at a color reduction of 50–70 % over 12 h was 45–55 %. Cammarota and Santanna (1992) developed a continuous packed bed bioreactor in which P. chrysosporium was immobilized on polyurethane foam particles. The bioreactor operation at a hydraulic retention time of 5–8 days was able to promote 70 % decolorization. The fungal biomass could be maintained in this process for at least 66 days without any appreciable loss of activity.

To apply the MyCOPOR process on an industrial scale, relatively big reactors (diameter, 70 and 100 mm; volume 4 to 16 l) were prepared and filled with polyurethane foam cubes (1 cm^3) as carrier material. Long-term experiments were successfully carried out and it was decided to build a small pilot reactor at a large paper mill in Austria (Jaklin-Farcher et al. 1992). However, many aspects related to the operating conditions must be further investigated and improved. A disadvantage of these treatment processes is that P. chrysosporium required high concentrations of oxygen and energy sources such as glucose or cellulose as well as various basal nutrients, mineral solution and Tween 80 (Messner et al. 1990). Kang et al. (1996) developed a submerged biofilter system in which mycelia of P. chrysosporium were attached to media (net ring type) and used to dispose waste water from a pulp mill. Maximum reduction of BOD, COD and lignin concentrations were 94, 91, and 90 %, respectively, in 12 h of hydraulic retention time.

Matsumoto et al. (1985) demonstrated that RBC treatment of extraction stage effluent was effective for the removal of AOX and color. Removal of AOX was determined to be 62, 43, and 45 % per day for the low molecular weight fraction of extraction stage effluent, high molecular weight fraction of the same, and unfractionated extraction stage effluent respectively. After further optimization, 49 % of the high molecular weight AOX was transformed to inorganic chloride in 1 day and 62 % in 2 days. The chloride concentration increased simultaneously with decreasing AOX including decolorization.

Singhal et al. (2005) studied treatment of pulp and paper mill effluent by *P. chlysosporium* at two different pHs, 5.5 and 8.5. At both pHs, color, COD, lignin content, and total phenols of the effluent significantly declined after bioremediation. However, greater decolorization and reduction in COD, lignin content, and total phenols were observed at pH 5.5. Such bioremediated effluent of pulp and paper mill could gainfully be utilized for crop irrigation.

Egyptian researchers applied the fungus, *P. chrysosporium* DSMZ 1556, to the microbiological processing of mill effluents (Abdel-Fattah et al. 2001). Experiments were conducted to compare the decolorization of paper mill effluents using this fungus under free cell, repeated batch, and coimmobilization systems. Immobilization and coimmobilization of the fungus was accomplished using alginate and activated charcoal. A two-fold increase in color reduction was achieved using fungus that was immobilized in alginate compared with alginate used alone as bioadsorbent. Similarly, a further 40 % increase in decolorization was found to occur with the cells co-immobilized with alginate and charcoal, compared with alginate and charcoal used alone. The results are ascribed to the ability of the immobilization and the protective barrier formed by the adsorbent to provide greater control over the remediation process.

Another white-rot fungus, *C. versicolor*, removed 60 % of the color of combined bleach Kraft effluents within 6 days in the presence of sucrose (Livernoche et al. 1983). Decolorization of the effluent was more efficient when the concentration of sucrose and inoculum was high. When the fungus was immobilized in calcium alginate gel, it removed 80 % color from the same effluent in 3 days in the presence of sucrose.

Biological reactors of the airlift type using calcium alginate beads to immobilize the fungus *C. versicolor* have been used to study the continuous decolorization of Kraft mill effluents (Royer et al. 1985). The effluent used contained only sucrose. An empirical kinetic model was proposed to describe the decolorization process caused by this fungus, but it did not shed any light on the chemical mechanism involved in the decolorization.

Bergbauer et al. (1991) showed that *C. versicolor* efficiently degraded chlorolignins. More than 50 % of the chlorolignins were degraded in 9 day incubation period, resulting in a 39 % reduction in AOX and 84 % decrease in effluent color. In a 3 l laboratory fermenter, with 0.8 % glucose and 12 mM ammonium sulfate, a color reduction of about 88 % was achieved in 3 days. Simultaneously, the AOX was reduced by 45 % in 2 days.

Direct use of suspended mycelium of *C. versicolor* may not be feasible because of the problem of viscosity, oxygen transfer, and recycling of the fungus. The fungus was therefore grown in the form of pellets, thus eliminating problems with biomass recycling and making it possible to use a larger amount (Royer et al. 1985). The rate of decolorization with fungal pellets was almost ten times as high in batch culture as in continuous culture under similar conditions (Royer et al. 1985).

Bajpai et al. (1993) reported 93 % color removal and 35 % COD reduction, from first extraction stage effluent with mycelial pellets of *C. versicolor* in 48 h in a batch reactor, whereas, in a continuous reactor, the same level of color and COD reduction was obtained in 38 h. No loss in decolorization ability of mycelial pellets was obtained when the reactor was operated continuously for more than 30 days. Mehna et al. (1995) also used *C. versicolor* for decolorization of effluents from a pulp mill using agriresidues. With an effluent of 18,500 color units, a color reduction of 88–92 % with COD reduction of 69–72 % was obtained. Royer et al. (1991) described the use of pellets of *C. versicolor* to decolorize ultra filtered Kraft liquor in nonsterile conditions with a negligible loss of activity. The rate of decolorization was observed to be linearly related to the liquor concentration and was lower than that obtained in the MyCoR process. Simple carbohydrates were found to be essential for effective decolorization with this fungus and a medium composed of inexpensive industrial by-products provided excellent growth and decolorization (Archibald et al. 1990).

Pallerla and Chambers (1996) have shown that immobilization of *T. versicolor* in urethane prepolymers leads to significant reductions in color and AOX in the treatment of bleach effluents. color reduction ranging of 72–80 % and AOX reduction of 52–59 % was found in a continuous bioreactor with a residence time of 24 h. The highest color removal rate of 1920 PCU per day was achieved at an initial color concentration of 2700 PCU. The biocatalyst remained intact and stable after an extended 32-day operation.

Treatment of extraction stage effluent with ozone and *C. versicolor* has also been tried (Roy-Arcand and Archibald 1991b). Both ozone treatment and biological treatment effectively destroyed effluent chromophores but the fungal process resulted in greater degradation as expressed by COD removal. Monoaromatic chlorophenolics and toxicity were removed partially by ozone and completely by *C. versicolor*. The combination of a brief ozone treatment with a subsequent fungal treatment revealed a synergism between the two decolorization mechanisms.

Pendroza et al. (2007) carried out experiments with *T. versicolor* to see the effect of using a sequential biological and photocatalytic treatment on COD, color removal and the degradation of chlorophenolic compounds in bleaching effluent. *T. versicolor* was cultured in an Erlenmeyer flask with wheat bran broth and 100 polyurethane foam cubes. The culture was incubated for 9 days at 25 °C. After treatment there was an 82 % reduction in COD and color, and significant reductions in chlorophenols. When this was followed by photolysis with titanium dioxide/Ru-Se, COD fell by 97 %, and there was a 92 % reduction in color and 99 % reduction in chlorophenols.

van Driessel and Christov (2001) studied the bioremediating abilities of *C. versicolor* and *Rhizomucor pusillus* applied to plant effluents in a RBC reactor. The decolorization was directly proportional to initial color intensities and its extent was not adversely affected by color intensity, except at the lowest level tested.

Decolorization of 53–73 % was obtained using a HRT of 23 h. With *R. pusillus*, 55 % of AOX were removed compared to 40 % by *C. versicolor*. Fungal treatment with both *R. pusillus* and *C. versicolor* rendered the effluent virtually nontoxic and the addition of glucose to the decolorization media stimulated color removal by *C. versicolor*, but not with *R. pusillus*.

White-rot fungus *C. subvermispora* has been found to decolorize, dechlorinatez and detoxify the pulp mill effluents at low cosubstrate concentration (Nagarathnamma et al. 1999). The fungus removed 91 % of the color and 45 % of the COD in 48 h under optimum conditions. The reductions in lignin, AOX, and EOX were 62, 32, and 36 %, respectively. The color removal rate was 3185 PCU/day at an initial color concentration of 7000 PCU. Monomeric chlorinated aromatic compounds were removed almost completely and toxicity to Zebra fish was eliminated.

Belém et al. (2008) used *Pleurotus sajor caju* and *P. ostreatus* to remove color from Kraft mill effluent by an ASB process. Absorbance reduction of 57 and 76 % was observed after 14 days of treatment with glucose by *P. sajor caju*, at 400 and 460 nm, respectively. Lower values of absorbance reduction were observed with additives and inoculated with the same species (22–29 %). Treatment with *P. ostreatus* was more efficient in the effluent with additives, 38.9–43.9 % of reduction. Higher growth rate of *P. sajor caju* was observed in the effluent with glucose. Biological treatment resulted in 65–67 % reduction of COD after 14 days revealing no differences for each effluent composition and inoculated species. Selvem et al. (2002) used *Fomes lividus* and *T. versicolor* to treat pulp and paper industry effluents. On the laboratory scale, a maximum decolorization of 63.9 % was achieved by *T. versicolor* on the fourth day. COD was reduced by 59.3 % by each of the two fungi. On the pilot scale, a maximum decolorization of 68 % was obtained on 6th day by *T. versicolor*. Ragunathan and Swaminathan (2004) studied the ability of *Pleurotus spp.* to treat pulp and paper mill effluent on a laboratory and pilot scale. *P. sajor-caju* decolorized the effluent by 60–66.7 % and COD by 57.2 %.

Shintani et al. (2002) used *Geotrichium candidum*, for the treatment of Kraft pulp bleaching effluent. With a glucose content of 30 g/L, a color removal of 78 % and a reduction in AOX of 43 % could be obtained after 1 week. Decolorization was not observed in the absence of added glucose. The average molecular weight of colored substances was reduced from 5,600 to below 3,000. color removal is believed to proceed via color adsorption to the cells followed by decomposition of the adsorbed materials.

Wu et al. (2005) explored the lignin-degrading capacity of five white-rot fungi—*P. chrysosporium*, *P. ostreatus*, *L. edodes*, *T. versicolor*, and S22. These fungi were grown on a porous plastic media, and were individually used to treat black liquor. Over 71 % of lignin and 48 % of COD were removed. Several factors, including pH, concentrations of carbon, nitrogen, and trace elements in wastewater, had significant effects on the degradation of lignin and the removal of COD.

Fukuzumi (1980) found a white-rot fungus, *Tinctoporia borbonica*, decolorizes Kraft waste liquor to a light yellow color. About 99 % color reduction was achieved after 4 days. Measurement of the culture filtrate by UV showed that the chlorine-oxylignin content also decreased with time and measurement of the culture filtrate plus mycelial extract after 14 days showed the total removal of the chlorine-oxylignin content.

Another white-rot fungus, *S. commune,* has also been found to decolorize and degrade lignin in pulp and paper mill effluent (Belsare and Prasad 1988). The addition of carbon and nitrogen not only improves the decolorizing efficiency, but also results in reduction of the BOD and COD of the effluent. Under optimum conditions, this fungus reduced the color of the effluent by 90 % and also reduced BOD and COD by 70 and 72 % during a two day incubation.

Duran et al. (1991) reported that preradiation of the effluent, followed by fungal culture filtrate treatment resulted in efficient decolorization. Moreover, when an effluent pre-irradiated in the presence of ZnO was treated with *L. edodes* (Esposito et al. 1991), a marked enhancement of the decolorization at 48 h was obtained (Duran et al. 1994).

The *zygomycete R. oryzae* has been reported to decolorize, dechlorinate, and detoxify extraction stage effluent at low cosubstrate concentrations. Optimum conditions for treatability were determined as pH 3–4.5 and temperature 25–40 °C (Nagarathnamma and Bajpai 1999). Under optimum conditions, the fungus removed 92–95 % color, 50 % COD, 72 % AOX, 37 % EOX, as well as all monoaromatic phenolics and toxicity. Significant reduction in chlorinated aromatic compounds was observed and toxicity to zebra fish was completely eliminated. The molecular weight of chlorolignins was substantially reduced after the fungal treatment. Another thermotolerant *zygomycete* strain, *R. pusillus* RM 7, could remove up to 71 % of color and substantially reduce COD, toxicity, and AOX levels in the effluent (Christov and Steyn 1998).

Kannan et al. (1990) reported about 80 % color removal and over 40 % BOD and COD reduction with the fungus *Aspergillus niger* in 2 days. Tono et al. (1968) reported that *Aspergillus sp.* and *Penicillium sp.* achieved 90 % decolorization in 1 week's treatment at 30 °C and at pH 6.8. Milstein et al. (1988) reported that these microorganisms removed appreciable levels of chlorophenols as well as chloroorganics from the bleach effluent. Gokcay and Taseli (1997) have reported over 50 % AOX and color removal from softwood bleach effluents in less than 2 days with *Penicillium sp.* Bergbauer et al. (1992) reported AOX reduction by 68 % and color reduction by 90 % in 5 days with the fungus *Stagonospora gigaspora.* Toxicity of the effluent was reduced significantly. Few marine fungi have been also reported to decolorize the bleach plant effluents (Raghukumar et al. 1996, 2008). With *Trichoderma* sp. under optimal conditions, total color and COD decreased by almost 85 and 25 %, respectively, after cultivation for 3 days (Prasad and Joyce 1991).

Malaviya and Rathore (2007) reported bioremediation of pulp and paper mill effluent by an immobilized fungal consortium for the first time. They immobilized two basidiomycetous fungi and a deuteromycetous fungus on nylon mesh and used the consortium for bioremediation of pulp and paper mill effluent in a continuously aerated bench-top bioreactor. The treatment resulted in the reduction of color, lignin, and COD of the effluent in the order of 78.6, 79.0, and 89.4 % in 4 days.

Singhal and Thakur (2009) took up geno-toxicity analysis along with effluent treatment to evaluate the efficiency of biological treatment process for safe disposal of treated effluent. Four fungi were isolated from sediments of pulp and paper mill in which PF4 reduced color (30 %) and lignin content (24 %) of the effluent on the third day. The fungal strain was identified as *Emericella nidulans*. Decolorization of effluent improved by 31 % with reduction in color (66.66 %) and lignin (37 %) after treatment by fungi in a shake flask. Variation in pH from 6 to 5 had the most significant effect on decolorization (71 %).

Chuphal et al. (2005) applied *Paecilomyces sp.* and *Pseudomonas syringae* for treatment of pulp and paper mill effluent in a two-step and three-step fixed film sequential bioreactor containing sand and gravel at the bottom of the reactor for immobilization of microbial cells. The microbes exhibited significant reduction in color (88.5 %), lignin (79.5 %), COD (87.2 %), and phenol (87.7 %) in two-step aerobic sequential bioreactor, and color (87.7 %), lignin (76.5 %), COD (83.9 %), and phenol (87.2 %) in three-step anaerobic–aerobic sequential bioreactor.

Sequential treatment using fungal process followed by photo-catalytical treatment on COD, color removal, degradation of chlorophenolic compounds in bleach effluent has been studied by Pendroza et al. (2007). The overall reduction was 97 % in COD, 92 % in color, and 99 % in chlorophenols.

Sorce Inc. uses a combination of fungi and facultative bacteria to degrade lignin. This technology was started up in a Southeastern Kraft pulp mill with a wastewater flow of 15 million gallons/day. The starting color value of the wastewater averaged 1880 Pt–Co units. After nutrient conditioning and application of the microorganisms to the mill's lagoon, color reduced by more than 50 % (Sorce Inc. 2003). This technology is called fungal/bacterial sequencing or biogeochemical cycling. It uses specific microorganisms and manipulates their life cycles so that their degrading activities can be predicted and harnessed in a beneficial way. In this case, white-rot fungi and its ability to secrete highly oxidative enzymes is used to fragment the lignin structure into smaller compounds. These fragments are mineralized into CO_2 and water with facultative anaerobic bacteria. Mineralization is accelerated with the use of co-metabolites. The latent phase of the fungal growth is a time of limited food, nutrients, or adverse environmental conditions that results in a decrease in the microbial population. The life cycle manipulation is the basis of the white-rot/facultative bacteria sequencing technology. In other words, keeping the fungi in a latent phase at the same time keeping facultative bacteria in a prolonged exponential growth phase. Under these conditions, the white-rot fungi secretes a tremendous amount of enzymes catalyzing the lignin degradation reaction that become the prime food source for the bacteria ultimately mineralizing the degraded lignin fragments and reducing color, toxicity, BOD, COD, etc. of the water to make it suitable for certain applications. This technology seems to be quite effective but requires large land area—tens of hectares.

A novel two-stage process using chemical hydrogenation as a first-stage treatment, followed by biological oxidation showed promise in substantially reducing the color of pulp mill effluents. In a pilot plant study using two 20 L reactors in series, the addition of sodium borohydride to the first reactor, for a residence time

Table 5.4 Comparison of color and AOX reduction by *P. chrysosporium* and *T.versicolor*

Fungus	Operation mode	Residence time (day)	Max. color reduction (%)	Max. AOX reduction (%)	Reference
P. chrysosporium					
Mycelium immobilized on rotating disc	Batch	1	90	–	Yin (1989)
T. versicolor					
Mycelial pellets immobilized in Ca-alginate	Batch	3	80	–	Livernoche et al. (1983)
Mycelial pellets	Batch	2	61	–	Royer et al. (1985)
Free cells	Batch	3	88	45	Bergbauer et al. (1991)
Mycelial pellets	Batch	5	80	–	Archibald et al. (1990)
Mycelial pellets	Batch	3	88	–	Mehna et al. (1995)
P. chrysosporium					
Mycelium immobilized on porous material	Continuous	0:5	60	55	Messner et al. (1990)
Mycelium immobilized on polyurethane foam	Continuous	5–8	70	–	Cammarota and Santanna (1992)
Mycelium immobilized on net ring type	Continuous	0.5	91	–	Kang et al. (1996)
T. versicolor					
Mycelial pellets immobilized in Ca- alginate	Continuous	0.7	45	–	Royer et al. (1983)
Mycelial pellets	Continuous	0.6–1.2	50	–	Royer et al. (1985)
Mycelial pellets	Continuous	1	78	42	Pallerla and Chambers (1995)
Mycelial pellets immobilized in Ca- alginate beads	Continuous	1	80	40	Pallerla and Chambers (1996)
Mycelial pellets	Continuous	1.6	93	–	Bajpai et al. (1993)

of 1 day, resulted in a 97 % reduction in color. Subsequent biological oxidation in the second reactor reduced BOD (99 %), COD (92 %), and TSS (97 %) (Ghoreishi and Haghighi 2007).

Table 5.4 shows the comparison of the results for color and AOX reduction by a *P. chrysosporium* and *T. versicolor.*

References

Abdel-Fattah YR, Hoda HH, El-Kassas HY, Sabry SA (2001) Bioprocess development of paper mill effluent's decolourization by *Phanerochaete chrysosporium* DSMZ 1556. Fresenius Environ Bull 10(10):761–765

Achoka JD (2002) The efficiency of oxidation ponds at the Kraft pulp and paper mill at Webuye in Kenya. Water Res 36:1203–1212

Afonso MD, Pinho MN (1991) Membrane separation processes in pulp and paper production. Filtr Sep 28(1):42–44

Akhtar M, Blanchette RA, Myers G (1998) An overview of biomechanical pulping research. In: Young RA, Akhtar M (eds) Environmentally friendly technologies for the pulp and paper industry. Wiley, New York, pp 309–340

Ali M, Sreekrishnan TR (2007) Anaerobic treatment of agricultural residue based pulp and paper mill effluents for AOX and COD reduction. Process Biochem 36(1–2):25–29

Altnbas U, Eroglu V (1997) Treatment of bleaching effluent in sequential activated sludge and nitrification systems. Fresen Envir Bull 6:103–108

Amat AM, Arques A, Lopez F, Miranda MA (2005) Solar photo-catalysis to remove paper mill wastewater pollutants. Sol Energy 79:393–401

Anderson JR, Amini B (1996) Hydrogen peroxide bleaching. In: Dence CW, Reeve DW (eds) Pulp bleaching: principles and practice. Tappi Press, Atlanta, p 411

Andreasan K, Agertved J, Petersen JO, Skaarup H (1999) Improvement of sludge settleability in activated sludge plants treating effluent from pulp and paper industries. Water Sci Technol 40(11–12):215–221

Archibald F, Paice MG, Jurasek L (1990) Decolourization of Kraft bleaching effluent chromophores by *Coriolus (Trametes) versicolour*. Enz Microbiol Technol 12:846–853

Asselin C, Collin D, Graff S (2000) Effluent treatment for chronic toxicity removal with the suspended carrier biofilm reactor. In; Tappi international environmental conference and exhibit. Denver CO, vol 2. Norcross, GA 30092, USA: Technical Association for Pulp and Paper Industry (TAPPI); May 2000, pp 805–811

Ataberk S, Gokcay CF (1997) Removal of chlorinated organics from pulping effluents by activated sludge process. Fersen Envir Bull 6:147–153

Bajpai P, Mehna A, Bajpai PK (1993) Decolourization of Kraft bleach effluent with white rot fungus *Trametes versicolour*. Process Biochem 28:377–384

Bajpai P, Ananad A, Bajpai PK (2006) Bleaching with lignin oxidizing enzymes. Biotechnol Ann Rev 12:349–378

Bajpai P (2012a) Biotechnology for pulp and paper processing. Springer Science+Business Media, New York, p 414

Bajpai P (2012b) Environmentally benign approaches for pulp bleaching, 2nd edn. Elsevier Science B.V, p 416

Barr TA, Taylor T, Duff S (1996) Effect of HRT, SRT and temperature on the performance of activated sludge reactors treating bleached mill effluent. Water Res 30(4):799–802

Begum JA, Menezes GB, Moo-Young HK (2012) Treatment of pulp and paper mill wastewater. Water Environ Res 84:1502–1510

Belém A, Panteleitchouk AV, Duarte AC, Rocha-Santos TAP, Freitas AC (2008) Treatment of the effluent from a Kraft bleach plant with white rot fungi. Pleurotus sajor caju and Pleurotus ostreatu Global NEST J 10(3):426–431

Belsare DK, Prasad DY (1988) Decolourization of effluent from the bagasse based pulp mills by white-rot fungus *Schizophyllum commune*. Appl Microbiol Biotech 28:301–304

Bergbauer M, Eggert C, Kraepelin G (1991) Degradation of chlorinated lignin compounds in a bleach effluent by the white-rot fungus *Trametes versicolour*. Appl Microbiol Biotechnol 35(1):105–109

Bergbauer M, Eggert C, Kalnowski G (1992) Biotreatment of pulp mill bleachery effluent with the Coelomycetous fungus *Stagonospora gigaspora*. Biotech Lett 14(4):317–322

Berube PR, Hall ER (2000) Fate and removal kinetics of contaminants contained in the evaporator condensate during treatment for reuse using a high temperature membrane bioreactor. In: Proceedings of the 86th PAPTAC annual meeting, Montreal, Quebec. Canada: Pulp and Paper Technical Association of Canada, p B67

Bollag JM, Shottleworth KL, Anderson DH (1988) Laccase-mediated detoxification of phenolic compounds. Appl Env Microbiol 54:3086–3091

Boman B, Frostell B, Ek M, Eriksson KE (1988) Some aspects on biological treatment of bleached pulp effluents. Nord Pulp Pap Res J 1:13–18

Borch-Due A, Anderson R, Opheim B (1997) Treatment of integrated newsprint mill wastewater in moving bed biofilm reactors. Water Sci Technol 35(2–3):173–180

Breed D, Shackford LD, Pereira ER, Colodette JL (1995) Cost-effective retrofit of existing bleach plants to ECF and TCF bleached pulp production using a novel peroxide bleaching process. Pulping Conference, Chicago, IL, USA, 1–5 Oct 1995, Book 2, pp 779–788

Brite CEC (1994) Nanofiltration joins electrodialysis in recycling pulp bleach effluents. Chemical Eng 101:23–28

Bryant CV, Amy GL, Allemen BC (1987) Organic halide and organic carbon distribution and removal in a pulp and paper wastewater lagoon. J Water Pollut Control Fed 59(10):890–896

Bryant CW, Barkley WA (1990) The capabilities of conventional treatment systems for removal of chlorinated organic compounds from pulp and paper wastewater. Pacific Paper EXPO Technical Conference Proceedings, Program 7: Environment, Vancouver, BC

Bryant CW, Amy GL, Neil R, Ahmed S (1988) Partitioning of organic chlorine between bulk water and benthal interstitial water through a Kraft mill aerated lagoon. Wat Sci Tech 20(1):73–79

Bryant CW, Avenell JJ, Barkley WA, Thut RN (1992) The removal of chlorinated organics from conventional pulp and paper wastewater treatment systems. Water Sci Technol 26(1–2):417–425

Bryant CW, Barkley WA, Garett RM, Gardner FD (1997) Biological nitrification of Kraft wastewater. Water Sci Technol 35(2–3):147–153

Buckley DB (1992) A review of pulp and paper industry experience with biological treatment process bacterial augmentation. In: Tappi environmental conference, Tappi Press Atlanta, GA. U.S.A, pp 750–810

Bullock JM, Bicho PA and Saddler JN (1994) The effect of high molecular weight organics in bleached Kraft mill effluent on the biological removal of chlorinated phenolics. In: Proceedings of 1994 environmental conference, pp 371–378

Call HP (1991) Laccases in delignification, bleaching and wastewater treatment. Patent No. DE 4137761

Cammarota MC, Santanna GL Jr (1992) Decolourization of Kraft bleach plant E_1 stage effluent in a fungal bioreactor. Environ Technol 13:65–71

Campbell AG, Gerrard ED, Joyce TW (1982) The MyCoR process for colour removal from bleach plant effluent: bench-scale studies. In: Proceedings of the Tappi research and development conference, North Carolina, Tappi Press, Atlanta, Ga, pp 209–214

Chandra R (2001) Microbial decolourisation of pulp mill effluent in presence of nitrogen and phosphorous by activated sludge process. J Environ Biol 22(1):23–27

Chandra R, Singh S, Krishna Reddy MM, Patel DK, Purohit HJ, Kapley A (2008) Isolation and characterization of bacterial strains Paenibacillus sp. and Bacillus sp. for Kraft lignin decolourization from pulp paper mill waste. J Gen Appl Microbiol 54(6):399–407

Chernoberezhskii YuM, Dyagileava AS, Barysheva IA (1994) Coagulation treatment of wastewaters from paper and pulp plants. Russ J Appl Chem 67(3):354–359

Chernysh A, Liss NS, Allen GD (1992) A batch study of the aerobic and anaerobic removal of chlorinated organic compounds in an aerated lagoon. Water Pollut Res J Can 27(3):621–638

Christov LP and Steyn MG (1998) Modifying the quality of a bleach effluent using Mucoralean and white-rot fungi. In: Proceedings of 7th international conference on biotechnology in the pulp and paper industry. Vancouver, Canada, pp C203–C206

Chuphal Y, Kumar V, Thakur IS (2005) Biodegradation and decolourization of pulp and paper mill effluent by anaerobic and aerobic microorganisms in a sequential bioreactor. World J Microbiol Biotechnol 21(8–9):1439–1445

Clark T, Bruce M, Anderson S (1994) Decolourization of E_1 stage bleach effluent by combined hypochlorite oxidation and anaerobic treatment. Wat Sci Technol 29(5/6):421–432

Das CP, Patnaik LN (2000) Removal of lignin by industrial solid wastes. Pract Period Hazard Toxic Radioact Waste Manag 4(4):156–161

Davis S, Burns RG (1992) Covalent immobilisation of laccase on activated carbon for phenolic effluent treatment. Appl Microbiol Biotechnol 37:474–479

De Pinho MN, Minhalma M, Rosa MJ, Taborda F (2000) Integration of flotation/ultrafiltration for treatment of bleached pulp effluent. Pulp Pap Can 104(4):50–54

Deardorff TL, Willhelm RR, Nonni AJ, Renard JJ, Phillips RB (1994) Formation of polychlorinated phenolic compounds during high chlorine dioxide substitution and bleaching. Tappi J 77(8):163–173

Deshpande SH, Khanolkar VD and Pudumjee KD (1991) Anaerobic-aerobic treatment of pulp mill effluents: a viable technological option. In: Proceedings of the international workshop on small scale chemical recovery, high yield pulping and effluent treatment, 16–20 Sept 1991, New Delhi, India, pp 201–213

Dezotti M, Innocentini-Mei LH, Duran N (1995) Silica immobilised enzyme catalysed removal of chlorolignins from Kraft effluent. J Biotechnol 43:161–167

Dilek FB, Goekcay CF (1994) Treatment of effluents from hemp based pulp and paper industry: I. Waste characterization and physicochemical treatability. Wat Sci Technol 29(9):161–163

Dorica J, Elliott A (1994) Contribution of non-biological mechanisms of AOX reduction attained in anaerobic treatment of peroxide bleached TMP mill effluent. In: Proceedings of Tappi international environmental conference, pp 157–163

Dube M, McLean R, MacLatchy D, Savage P (2000) Reverse osmosis treatment: effects on effluent quality. Pulp Pap Can 101(8):42–45

Duran N, Dezotti M, Rodriguez J (1991) Biomass photochemistry—XV: photobleaching and biobleaching of Kraft effluent. J Photochem Photobiol 62:269–279

Duran N, Esposito E, Innocentini-Mei LH, Canhos VP (1994) A new alternative process for Kraft E_1 effluent treatment, a combination of photochemical and biological methods. Biodegradation 5(1):13–19

Eaton DC, Chang HM, Kirk TK (1980) Fungal decolourization of Kraft bleach plant effluents. Tappi J 63(10):103–106

Eaton DC, Chang HM, Joyce TW, Jeffries TW, Kirk TK (1982) Method obtains fungal reduction of the colour of extraction-stage Kraft bleach effluents. Tappi J 65(6):89–92

Edeline F, Lambert G, Fatliccioni H (1988) Anaerobic treatment of mixed hardwood Kraft pulp cooking condensates with first alkaline stage effluents. In: G Grassi (ed) Energy from Biomass–4. Proceedings of the 3rd contractors meeting, pp 463–475

Eeckhaut M, Alaerts G, Pipyn P (1986) Anaerobic treatment of paper mill effluents using polyurethane foam carriers reactor (PCR) technology. In: PIRA paper and board division seminar. Cost effective treatment of paper mill effluents using anaerobic technologies, 14–15 Jan 1986, Leatherhead, UK

EK M, Eriksson KE (1987) External treatment of bleach plant effluent. In: 4th international symposium on wood and pulping chemistry, Paris

EK M, Kolar MC (1989) Reduction of AOX in bleach plant effluents by a combination of ultrafiltration and biological methods. In: Proceedings of 4th international biotechnology, conference in pulp and paper industry, Raleigh, North Carolina, 16–19 May 1989, pp 271–278

Eriksson KE (1990) Biotechnology in the pulp and paper industry. Water Sci Technol 24:79–101

Eriksson KE, Kolar MC (1985) Studies on microbial and chemical conversions of chlorolignins. Environ Sci Technol 19(12):1219–1224

Esposito E, Canhos VP, Duran N (1991) Screening of lignin degrading fungi for removal of colour from Kraft mill wastewater with no additional extra carbon source. Biotech Lett 13(8):571–576

Evans T, Sweet W, Monlescu DJ (1994) Applying proven technology to eliminate Kraft bleach plant effluents. In: Proceedings of annual meeting technical section, Canadian Pulp and Paper Association, Montreal, Canada, pp 237–242

Falth F (2000) Ultrafiltration of E1 stage effluent for partial closure of the bleach plant. In: Proceedings of 86th PAPTAC annual meeting, Montreal, Quebec. Canada: Pulp and Paper Technical Association of Canada, p B85

Farrel RL (1987a) Use of rLDM™ 1-6 and othe ligninolytic enzymes. WO 87/00564

Farrel RL (1987b) Industrial applications of lignin transforming enzymes. Philos Trans R Soc London A 321:549–553

Ferguson JF, Dalentoft E (1991) Investigation of anaerobic removal and degradation of organic chlorine from Kraft bleaching wastewaters. Wat Sci Tech 24:241–250

Ferguson JF, Luonsi A, Ritter D (1990) Sequential anaerobic/aerobic biological treatment of bleaching wastewaters. In: Proceedings of Tappi environmental conference. Atlanta, GA, pp 333–338

Ferrer I, Dezotti M, Duran N (1991) Decolourization of Kraft effluent by free and immobilized lignin peroxidase and horseradish peroxidase. Biotech Lett 13:577–582

Field JA (1986) Method for biological treatment of waste waters containing nondegradable phenolic compounds and degradable nonphenolic compounds. EP Patent 1986; EP 238148

Field JA, Lettinga G (1991) Treatment and detoxification of aqueous spruce bark extracts by *Aspergillus niger*. Wat Sci Technol 24:127–137

Field JA, Leyendeckers MJH, Sierra-Alvarez R, lettinga G, Habets LHA (1988) The methanogenic toxicity of bark tannins and the anaerobic biodegradability of water soluble bark matter. Wat Sci Tech 20:219–240

Field JA, Leyendeckers MJH, Sierra–Alvarez R, Lettinga G (1991) Continuous anaerobic treatment of auto-oxidized bark extracts in laboratory-scale columns. Biotechnol Bioeng 37:247–255

Fitzsimonas R, Ek M, Eriksson K-EL (1990) Anaerobic dechlorination/degradation of chlorinated organic compounds of different molecular masses in bleach plant effluents. Environ Sci Tech 24:1744–1748

Forss K, Jokinen K, Savolainen M, Williamson H (1987) Utilization of enzymes for effluent treatment in the pulp and paper industry. In: Proceeding of the 4th international symposium on wood and pulping chemistry, vol 1, Paris, France, pp 179–183

Franta JR, Wilderer PA (1997) Biological treatment of papermill wastewater by sequencing batch reactor technology to reduce residual organics. Water Sci Technol 35(1):129–136

Franta J, Helmreich B, Pribyl M, Adamietz E, Wilderer PA (1994) Advanced biological treatment of papermill wastewaters; effects of operating conditions on COD removal and production of soluble organic compounds in activated sludge systems. Water Sci Technol 30(3):199–207

Freire RS, Kunz A, Duran N (2000) Some chemical and toxicological aspects about paper mill effluent treatment with ozone. Environ Technol 21:717–721

Fujita K, Kondo R, Sakai K, Kashino Y, Nishida T, Takahara Y (1991) Biobleaching of Kraft pulp using white-rot fungus IZU-154. Tappi J 74(11):123–127

Fukuzumi T (1980) Microbial decolourization and defoaming of pulping waste liquor. In: Kirk TK, Chang HM, Higuchi T (eds) Lignin biodegradation: microbiology, chemistry and potential applications, 2nd edn. CRC Press, Boca Raton, pp 161–171

Fukuzumi T, Nishida A, Aoshima K, Minami K (1977) Decolourization of Kraft waste liquor with white-rot fungi-I: Screening of the fungi and culturing condition for decolourization of Kraft waste liquor. Mokuzai Gakkaishi 23(6):290–298

Fulthorpe RR, Allen DG (1995) A comparison of organochlorine removal from bleached Kraft pulp and paper mill effluents by dehalogenating *Pseudomonas, Ancylobacter* and *Methylobacterium* strains. Appl Microbiol Biotechonol 42:782–789

Galeno G, Agasin E (1990) Screening of white-rot fungi for efficient decolourization of bleach pulp effluents. Biotechnol Lett 12(11):869–872

Ganjidoust H, Tatsumi K, Yamagishi T, Gohlian R (1996) Effect of synthetic and natural coagulant on lignin removal from pulp and paper wastewater. In: Proceedings of 5th IAWQ symposium for industrial wastewaters, Vancouver, Canada, p 305

Ganjidoust H, Tatsumi K, Yamagishi T, Gholian RN (1997) Effect of synthetic and natural coagulant on lignin removal from pulp and paper waste water. Water Sci Technol 35(2–3):291–296

Gergov M, Priha M, Talka E, Valtilla O, Kangas A, Kukkonen K (1988) Chlorinated organic compounds in effluent treatment at Kraft mills. Tappi J 71(12):175–184

Ghoreishi SM, Haghighi MR (2007) Chromophores removal in pulp and paper mill effluent via hydrogenation-biological batch reactors. Chem Eng J 127(1–3):59–70

Gokcay CF, Taseli BK (1997) Biological treatability of pulping effluents by using a *Penicillium* sp. Fresenus Envir Bull 6:220–226

Goronzy M, Demoulin G, Jager A, Srebotnik E, Messener K (1996) The use of the cyclic activated treatment technology for wastewater treatment in the pulp and paper industry. In: Proceedings 6th international conference biotechnology in pulp and paper industry, pp 239–245

Graves JW, Joyce TW, Jameel H (1993) Effect of chlorine dioxide substitution, oxygen delignification and biological treatment on bleach plant effluent. Tappi J 76(7):153–159

Guo HY, Chang HM, Joyce TW, Glasser JH (1990) Degradation of chlorinated phenols and guaiacols by the white rot fungus *Phanerochaete chrysosporium*. In: Kirk TK, Chang HM (eds) Biotechnology in pulp and paper manufacture. Butterworth-Heinemman, Stoneham, pp 223–230

Habets LHA, de Vegt AL (1991) Anaerobic treatment of bleached TMP and CTMP effluent in Biopaq UASB system. Wat Sci Tech 24:331–345

Habets LHA, Tielboard MH, Ferguson AMD, Prong CF, Chmelauskas AJ (1985) Onsite high rate UASB anaerobic demonstration plant treatment of NSSC waste water. Wat Sci Technol 20:87–97

Haggblom M, Salkinoja-Salonen M (1991) Biodegradability of chlorinated organic compounds in pulp bleaching effluents. Water Sci Technol (GB) 24(3/4):161–170

Hakulinen R (1982) The Enso-Fenox process for the treatment of Kraft pulp bleaching effluent and other waste waters of the forest industry. Paperi Ja Puu-Paper Och Tra 5:341–354

Hakulinen R (1988) The use of enzymes in the waste water treatment of pulp and paper industry- a new possibility. Water Sci Tech 20(1):251–262

Hall ER, Robson RD, Prong CF, Chmelauskas AJ (1986) Evaluation of anaerobic treatment for NSSC wastewater. In: Proceedings of the Tappi environmental conference, Atlanta, GA, pp 207–217

Hammel KE, Tardone PJ (1988) The oxidative 4-Dechlorination of polychlorinated phenols is catalyzed by extracellular fungal lignin peroxidases. Biochemistry 27:6563–6568

Hansen E, Zadura L, Frankowski S, Wachowicz M (1999) Upgrading of an activated sludge plant with floating biofilm carriers at Frantschach Swiecie S.A. to meet the new demands of year 2000. Water Sci Technol 40(11–12):207–214

Hirvonen A, Tuhkanen T, Kalliokoski P (1996) Formation of chlorinated acetic acids during UV/ H_2O_2-oxidation of ground water contaminated with chlorinated ethylenes. Water Sci Technol 32:1091–1102

Hostachy J, Lachenal D, Coste, G (1996) Ozonation of pulping bleaching effluents to reduce polluting charge. In: Proceedings of the Tappi international environmental conference, Orlando, Fla, pp 817–822

Huynh VB, Chang HM, Joyce TW, Kirk TK (1985) Dechlorination of chloroorganics by whiterot fungus. Tappi J 68(7):98–102

Jaklin-Farcher S, Szeker E, Stifter U, Messner K (1992) Scale up of the MYCOPOR reactor. In: Kuwahara M, Shimada M (eds) Biotechnology in pulp and paper industry. Tokyo, Japan, pp 81–85

Johnson T, Chatterjee A (1995) Activated sludge and surface aerators treat combined CTMP and Kraft effluent. Pulp Pap Can 96(8):26–29

Jokela JK, Laine M, EK M, Salkinoja-Salonen M (1993) Effect of biological treatment on halogenated organics in bleached Kraft pulp mill effluents studied by molecular weight distribution analysis. Environ Sci Technol 27(3):547–552

Jonsson AS, Jonsson C, Teppler M, Tomani P, Wannstrom S (1996) Treatments of paper coating colour effluents by membrane filtration. Desalination 105:263–276

Joyce TW, Pellinen J (1990) White rot fungi for the treatment of pulp and paper industry waste water. In: Tappi environmental conference, Seattle, 9–11 Apr 1990

Junna J, Ruonala S (1991) Trends and guidelines in water pollution control in the Finnish pulp and paper industry. Tappi J 74(7):105–111

Jurgensen SJ, Benjamin MM, Ferguson JF (1985) Treatability of thermomechanical pulping process effluents with anaerobic biological reactor. In: Proceedings Tappi environmental conference. Tappi Press, Atlanta, pp 83–92

Kamwilaisak K, Wright PC (2012) Investigating laccase and titanium dioxide for lignin degradation. Energy Fuels 26:2400–2406

Kang CH, On HK, Won CH (1996) Studies on the treatment of paper mill wastewater by *Phanerochaete chrysosporium*. In: Srebotnik E, Messener K (eds) Biotechnology in pulp and paper industry. Facultas-Universitatsverlag, Vienna, pp 263–266

Kannan K (1990) Decolourization of pulp and paper mill effluent by *Aspergillus niger*. World J Microbiol Biotechnol 6(2):114–116

Karimi S, Abdulkhani A, Karimi A, Ghazali AB, Ahmadun FR (2010) The effect of combination enzymatic and advanced oxidation process treatments on the colour of pulp and paper mill effluent. Environ Technol 31(4):347–356

Kennedy KJ, Graham B, Droste RL, Fernandes L, Narbaitz R (2000) Microtox and Ceriodaphnia dubia toxicity of BKME with powdered activated carbon treatment. Water SA 26(2):205–216

Klibanev AM, Morris ED (1981) Horseradish peroxidase for the removal of carcinogenic aromatic amines from water. Enz Microb Technol 3:119–122

Knudsen L, Pedersen JA, Munck J (1994) Advanced treatment of paper mill effluents by a two-stage activated sludge process. Water Sci Technol 30(3):173–181

Korhonen S, Tuhkanen T (2000) Effects of ozone on resin acids in thermomechanical pulp and paper mill circulation waters. Ozone: Sci Eng 22(6):575–584

Kortekaas S, Wijngaarde RR, Klomp JW, Lettinga G, Field JA (1998) Anaerobic treatment of hemp thermomechanical pulping wastewater. Wat Res 32(11):3362–3370

Lachenal D, Joncourt MJ, Froment P, Suty H (1996) Reduction of formation of AOX during chlorine dioxide bleaching. In: Proceedings of the international pulp bleaching conference, Washington, USA, pp 417–420

Lafond RA, Ferguson JF (1991) Anaerobic and aerobic biological treatment processes for removal of chlorinated organics from Kraft bleaching wastes. In: Proceedings of the Tappi environmental conference. Tappi press, Atlanta, GA, USA, pp 797–812

Lankinen VP, Inkeroinen MM, Pellinen J, Hatakka Al (1991) The onset of lignin modifying enzymes; decrease of AOX and colour removal by white rot fungi grown on bleach plant effluents. Wat Sci Tech 24(3/4):189–198

Leach JM, Mueller JC, Walden CC (1978) Biological detoxication of pulp mill effluents. Process Biochem 13(1):18–26

Lee SH, Kondo R, Sakai K (1994) Treatment of Kraft bleaching effluents by lignin degrading fungi-III: treatment by newly found fungus KS-62 without additional nutrients. Mokuzai Gakkaishi 40(6):612–619

Lee SH, Kondo R, Sakai K, Sonomoto K (1995) Treatment of Kraft bleaching effluents by lignin-degrading fungi V: successive treatments with immobilized mycelium of the fungus KS-62. Mokuzai Gakkaishi 41(1):63–68

Leponiemi A (2008) Non-wood pulping possibilities-a challenge for the chemical pulping industry. Appita J 61(3):234–243

Lettinga G, Field JA, Sierra-Alvarez R, Vanlier JB, Rintala J (1990) Future perspective for the anaerobic treatment of forest industry wastewaters. Wat Sci Tech 24:91–102

Lindstrom K, Mohamed M (1988) Selective removal of chlorinated organics from Kraft mill effluents in aerated lagoons. Nordic Pulp Paper Res J 3:26–33

Liu HW, Lo SN, Lavallee HC (1996) Mechanisms of removing resin and fatty acids in CTMP effluent during aerobic biological treatment. Tappi J 79(5):145–154

Liu HW, Liss SN, Allen D (1997) Influence of anoxic conditioning of sludge on enhanced AOX removal in aerobic biological treatment systems. Wat Sci Tech 35(2):77–82

Livernoche D, Jurasek L, Desrochers M, Dorica J, Veliky IA (1983) Removal of colour from Kraft mill waste waters with cultures of white-rot fungi and immobilised mycelium of *Coriolous versicolour*. Biotech Bioeng 25:2055–2065

Lunan WE, Harden C, Krupa K (1995) Pilot trials of trickling filter for treatment of waste water from a newsprint mill. In: Proceedings of the Tappi environmental conference, Atlanta, GA, USA, pp 241–247

Lyr VH (1963) Enzymic detoxification of chlorinated phenols. Phytopathol 47:73–83

MacLean B, de Vegt A, Droste RL (1990) Role of resin acids in the anaerobic toxicity of Chemithermomechanical pulp waste water. Wat Res 24:1401–1405

Magnus E, Carlberg GE, Norske HH (2000a) TMP wastewater treatment including a biological high-efficiency compact reactor. Nord Pulp Pap Res J 15(1):29–36

Magnus E, Carlberg GE, Norske HH (2000b) TMP wastewater treatment including a biological high-efficiency compact reactor. Nord Pulp Pap Res J 5(1):37–45

Magnus E, Hoel H, Carlberg GE (2000c) Treatment of an NSSC effluent in a biological high-efficiency compact reactor. Tappi J 83(1):149–156

Malaviya P, Rathore VS (2007) Bioremediation of pulp and paper mill effluent by a novel fungal consortium isolated from polluted soil. Bioresour Technol 98(18):3647–3651

Mapple GE, Ambady R, Caron JR, Stralton SC, Vega Canovas RE (1994a) BFR™: a new process towards bleach plant closure. Tappi J 77(11):71–80

Mapple GE, Ambady R, Caron JR, Stralton SC, Vega Canovas RE (1994b) BFR: a new process towards bleach plant closure. In: Proceedings of international pulp bleaching conference. Atlanta, Ga, pp 253–258

Mathys RG, Branion RMR, Lo KV (1993) CTMP waste water treatment using rotating biological contactor. In: Proceedings of the 79th CPPA annual proceedings, pp 370–374

Mathys RG, Branion RMR, Lo KV, Anderson KB, Leyen P, Louie D (1997) CTMP waste water treatment using rotating biological contactor. Wat Qual Res J Can 32:771–774

Matsumoto Y, Yin CF, Chang HM, Joyce TW, Kirk TK (1985) Degradation of chlorinated lignin and chlorinated organics by a white rot fungus. In: Proceedings of 3rd ISWPC, Vancouver, pp 45–53

McCarthy PJ, Kennedy KJ, Droste RL (1990) Role of resin acids in the anaerobic toxicity of chemithermomechanical pulp wastewater. Wat Res 24:1401–1405

McCubbin N (1983) The basic technology of the pulp and paper industry and its environment protection practices. Report EPS 6-EP-83-1, Environment Canada, Ottawa

McDonough TJ (1995) Recent advances in bleached chemical pulp manufacturing technology Part 1. Tappi J 78(3):55–62

McLeay DJ (1987) Aquatic toxicity of pulp paper mill effluent: a review. Environment Canada Report EPS 4/PF/1 Beauregard Press Ltd, Ottawa Ontario, Canada

Mehna A, Bajpai P, Bajpai PK (1995) Studies on decolourization of effluent from a small pulp mill utilizing agriresidues with *Trametes versicolour*. Enz Microb Technol 17(1):18–22

Melcer H, Steel P, Mckinley A, Cook CR (1995) The removal of toxic contaminants from bleached Kraft mill waste water with enhanced activated sludge treatment. In: Proceedings of Tappi international environmental conference, pp 795–807

Merrill DT, Maltby CV, Kahmark K, Gerhardt M, Melecer H (2001) Evaluating treatment process to reduce metals concentrations in pulp and paper mill wastewaters to extremely low values. Tappi J 84(4):52

Messner K, Ertler G, Jaklin-Farcher S, Boskovsky P, Regensperger V, Blaha A (1990) Treatment of bleach plant effluents by Mycor system. In: Kirk TK, Chang HM (eds) Biotechnology in pulp and paper manufacture. Butterworth-Heinemann, Newton, pp 245–251

Milet GM, Duff SJB (1998) Treatment of Kraft condensates in a feedback controlled sequencing batch reactor. Water Sci Technol 38(4–5):263–271

Milstein O, Haars A, Majcherczyk A, Trojanowski J, Tautz D, Zanker H, Huttermann A (1988) Removal of chlorophenols and chlorolignins from bleaching effluents by combined chemical and biological treatment. Wat Sci Tech 20(1):161–170

Milstein O, Hwars A, Krause F, Huettermann xx (1991) Decrease of pollutant level of bleaching effluents and winning valuable products by successive flocculation and microbial growth. Wat Sci Technol 24(3/4):199–206

Mishra M, Thakur IS (2010) Isolation and characterization of alkalotolerant bacteria and optimization of process parameters for decolourization and detoxification of pulp and paper mill effluent by Taguchi approach. Biodegradation 21(6):967–978

Mittar D, Khanna PK, Marwaha SS, Kennedy JF (1992) Biobleaching of pulp and paper mill effluents by *Phanerochaete chrysosporium*. J Chem Technol Biotechnol 53:81–92

Mohamed M, Matayun M, Lim TS (1989) Chlorinated organics in tropical hardwood Kraft pulp and paper mill effluents and their elimination in an activated sludge treatment system. Pertanika 2(3):387–394

Monje PG, González-García S, Moldes D, Vidal T, Romero J, Moreira MT, Feijoo G (2010) Biodegradability of Kraft mill TCF biobleaching effluents: application of enzymatic laccase-mediator system. Water Res 44(7):2211–2220

Mortha G, Mckay LR, Cadel F, Rouger J (1991) AOX reduction in an activated sludge treatment of Kraft bleaching effluent. 6th ISWPC, pp 1–2

Murthy BSA, Sihorwala TA, Tilwankar HV, Killedar DJ (1991) Removal of colour from pulp and paper mill effluents by sorption technique: a case study. Indian J Environ Prot 11(5):360

Muurinen E (2000) Organosolv pulping: a review and distillation study related to peroxyacid pulping. Department of Process Engineering, University of Oulu, FIN-90014 University of Oulu, Finland, p 314

Nagarathnamma R, Bajpai P, Bajpai PK (1999a) Studies on decolourization, degradation and detoxification of chlorinated lignin compounds in Kraft bleaching effluents by *Ceriporiopsis subvermispora*. Process Biochem 34:939–948

Nagarathnamma R, Bajpai P, Bajpai PK (1999b) Decolourization of extraction stage effluent from chlorine bleaching of Kraft pulp by *Rhizopus oryzae*. Appl Environ Microbiol 65(3):1078–1082

Nancy SJ, Norman JC, Vandenbusch MB (1996) Removing colour and chlorinated organics from pulp mill bleach plant effluents by use of fly ash. Conserv Recycl 10(4):279–299

Narbaitz RM, Droste RL, Fernandes L, Kennedy KJ, Ball D (1997) PACT™ process for treatment of Kraft mill effluent. Water Sci Technol 35(2–3):283–290

Nishida T, Katagiri N, Tsutsumi Y (1995) New analysis of lignin-degrading enzymes related to biobleaching of Kraft pulp by white-rot fungi. In: 6th international conference on biotechnology in the pulp and paper industry, Vienna, Austria, 11–15 June 1995

Oeller HJ, Daniel I, Weinberger G (1997) Reduction in residual COD in biologically treated paper mill effluents by means of combined Ozone and Ozone/UV reactor stages. Water Sci Technol 35(2–3):269–276

Olofsson A (1996) Domsjo heats up Ornskoldsvik with biogas. Svensk Papperstidin 99(11):33–34

Paice MG (1995) Activated sludge treatment of mechanical pulp mill effluents containing sulfite. In: Proceedings of CPPA environmental conference, pp 81–86

Paice MG, Jurasek L (1984) Peroxidase catalyzed colour removal from bleach plant effluent. Biotechnol Bioeng 1984(26):477–480

Paice M, Elloitt A, Young F, Dorica J (1996) Activated sludge treatment of mechanical pulp mill effluents containing sulfites. Pulp Pap Can 97(9):88–92

Pallerla S, Chambers RP (1995) Continuous decolourization and AOX reduction of bleach plant effluents by free and immobilized *Trametes versicolour*. J Env Sci Health A30(2):432–437

Pallerla S, Chambers RP (1996) New urethane prepolymer immobilized fungal bioreactor for decolourization and dechlorination of Kraft bleach effluents. Tappi J 79(5):156–161

Parker WJ, Eric R, Farguhar GJ (1993a) Assessment of design and operating parameters for high rate anaerobic fermentation of segregated Kraft mill bleach plant effluents. Wat Environ Res 65(3):264–270

Parker WJ, Eric R, Farguhar GJ (1993b) Removal of chlorophenolics and toxicity during high rate anaerobic treatment of Kraft mill bleach plant effluents. Environ Sci Technol 27(9):1783–1789

Parthasarthy VR, Rudie GF, Mendiratta SK, Cawifield DW, Bryan HP (1994) Gaseous chlorine dioxide delignification of softwood pulps for high brightness and ultra low AOX. In: Proceedings of pulping conference. San Diego, California, USA, pp 1031–1038

Pejot F, Pelayo JM (1993) Colour and COD removals by ultrafiltration from paper mill effluents: semi industrial pilot test results. In: Proceedings of Tappi environmental conference. Atlanta, Ga, USA, pp 389–394

Pellinen J, Joyce TW, Chang HM (1988a) Dechlorination of high molecular weight chlorolignin by white rot fungus *Phanerochaete chrysosporium*. Tappi J 71(9):191–194

Pellinen J, Yin CF, Joyce TW, Chang HM (1988b) Treatment of chlorine bleaching effluent using a white-rot fungus. J Biotechnol 8(1):67–76

Pendroza AM, Mosqueda R, Alonso-Vante N, Rodriguez-Vazquez R (2007) Sequential treatment via Trametes versicolour and UV/titanium dioxide/RuxSey to reduce contaminants in waste water resulting from the bleaching process during paper production. Chemosphere 67(4):793–801

Pereira ER, Colodette JL, Barna JM (1995) Peroxide bleaching in pressurised equipment: influence of temperature and consistency. Papel 56(12):61–67

Pertulla M, Konrusdottin M, Pere J, Kristjansson JK, Viikari L (1991) Removal of acetate from NSSC sulphite pulp mill condensates using thermophilic bacteria. Wat Res 25:599–604

Pichon M, Rouger J, Junet E (1988) Anaerobic treatment of sulphur containing effluents. Wat Sci Technol 20:133–141

Pokhrel D, Viraraghavan T (2004) Treatment of pulp and paper mill wastewater: a review. Sci Total Environ 333(1–3):37–58

Poppius-Levlin K, Mustonen R, Huovila T, Sundquist J (1991) Milox pulping with acetic-acid. Paperi ja Puu-Paper Timber 73(2):154–158

Prasad DY, Joyce TW (1991) Colour removal from Kraft bleach plant effluents by *Trichoderma* sp. Tappi J 74(1):165–169

Prouty AL (1990) Bench scale development and evaluation of a fungal bioreactor for colour removal from bleach effluents. Appl Microbiol Biotechnol 32:490–493

Raghukumar C, Chandramohan D, Michael FC Jr, Reddy CA (1996) Degradation of lignin and decolourization of paper mill bleach plant effluent by marine fungi. Biotech Lett 18(1):105–106

Raghukumar C, D'Souza-Ticlo D, Verma AK (2008) Treatment of coloured effluents with lignin-degrading enzymes: an emerging role of marine-derived fungi. Crit Rev Microbiol 34:189–206

Raghuveer S, Sastry CA (1991) Biological treatment of pulp mill wastewater mand study of biokinetic constants. Indian J Environ Prot 11(8):614–621

Ragunathan R, Swaminathan K (2004) Biological treatment of a pulp and paper industry effluent by *Pleurotus* spp. World J Microbiol Biotechnol 20(4):389–393

Rai A, Krishna Reddy MM, Chandra R (2007) Decolourisation and treatment of pulp and paper mill effluent by lignin-degrading Bacillus sp. J Chem Technol Biotechnol 82(4):399–406

Raizer-Neto E, Pichon M, Beniamin MM (1991) Decreasing chlorinated organics in bleaching effluents in an anaerobic fixed bed reactor. In: Kirk TK, Chang HM (eds) Biotechnology in pulp and paper manufacture. Butterworth-Heinmann, Stoneham, pp 271–278

Ramaswammy V (1987) Biotechnology application in waste utilization and pollution abatement IPPTA convention issue. Indian Pulp and Paper Association, Saharanpur India

Reid ID, Bourbolnnais R, Paice MG (2010) Biopulping and biobleaching. In: Heitner C, Dimmel DR, Schmidt JA (eds) Lignin and lignans: advances in chemistry, Chap 15, pp 521–554

Reid TK, Simon A (2000) Feasibility study of sequencing batch reactor technology treating high strength foul condensate for methanol reduction. In: Tappi international environmental conference and exhibit, vol 1, pp 185–91, Denver, CO

Rempel W, Turk O, Sikes JEG (1990) Side by side activated sludge pilot plant investigations focussing on organochlorines. CPPA Spring Conference Jasper, Alberta

Richardson DA, Andras E, Kennedy KJ (1991) Anaerobic toxicity of fines in Chemithermomechanical pulp wastewaters: a batch assay-reactor study comparison. Wat Sci Tech 24:103–112

Rintala J, Lepisto S (1992) Anaerobic treatment of thermomechanical pulping wastewater at 35–70° C. Wat Res 26:1297–1305

Rintala J, Vuoriranta P (1988) Anaerobic-aerobic treatment of pulping effluents. Tappi J 71:201–207
Rinzema A, Lettinga G (1988) Anaerobic treatment of sulfate containing wastewater. Biotreatment Syst 3:65–109
Rogers IH, Davis JC, Kruzynski GM, Mahood HW, Servizi JA, Gordon JW (1975) Fish toxicants in Kraft effluents. Tappi J 58(7):136–140
Rohella RS, Choudhury S, Manthan M, Murthy JS (2001) Removal of colour and turbidity in pulp and paper mill effluents using polyelectrolytes. Indian J Environ Health 43(4):159–163
Roy-Arcand L, Archibald FS (1991a) Direct dechlorination of chlorophenolic compounds by laccases from *Trametes (Coriolus) versicolour*. Enzyme Microb Technol 13(3):194–203
Roy-Arcand L, Archibald FS (1991b) Comparison and combination of ozone and fungal treatments of a Kraft bleaching effluents. Tappi J 74(9):211–218
Royer G, Desrochers M, Jurasek L, Rouleau D, Mayer RC (1985) Batch and continuous decolourization of bleached Kraft effluents by white-rot fungus. J Chem Technol Biotechnol 35B:14–22
Royer G, Livernoche D, Desrochers M, Jurasek L, Rouleau D, Mayer RC (1983) Decolourization of Kraft mill effluent: kinetics of a continuous process using immobilized *Coriolus versicolour*. Biotech Lett 5:321–326
Royer G, Yerushalmi L, Rouleau D, Desrochers M (1991) Continuous decolourization of bleached Kraft effluents by *Coriolus versicolour* in the form of pellets. J Indl Microbiol 7:269–278
Ruggiero P, Sarkar JM, Bollag JM (1989) Detoxification of 2,4-dichlorophenol by a laccase immobilized on soil or clay. Soil Sci 147:361–370
Salkinoja–Salonen M (1990) Biodegradability and ecological considerations of organochlorine from bleached Kraft pulp mill effluent. In: 6th Colloquium on pulp and paper mill effluents, Toranto
Sarner E (1988) Anaerobic treatment of a mixture of condensate and caustic extraction liquor from dissolving pulp mill. Wat Sci Tech 20(1):279–281
Saunamaki R (1989) Biological waste water treatment in the Finnish pulp and paper industry. Pap Puu 2:158–164
Saunamaki R, Jokinen K, Jarvinen R, Savolainen M (1991) Factors affecting the removal and discharge of organic chlorine compound at activated sludge treatment plants. Wat Sci Technol 24(3/4):295–308
Schnell A, Steel P, Melcer H, Hodson PV, Carey JH (2000) Enhanced biological treatment of bleached Kraft mill effluents: I. Removal of chlorinated organic compounds and toxicity. Water Res 34(2):493–500
Selvam K, Swaminathan K, Song MH, Chae KS (2002) Biological treatment of a pulp and paper industry effluent by Fomes lividus and Trametes versicolour. World J Microbiol Biotechnol 18(6):523–526
Shawwa AR, Smith DW, Sego DC (2001) Color and chlorinated organics removal from pulp wastewater using activated petroleum coke. Water Res 35(3):745–749
Sheela V, Distidar MG (1989) Treatment of black liquor wastes from small paper mills. Indian J Environ Prot 9(9):661–666
Shere SM, Daly PG (1982) High rate biological treatment of TMP effluent. Pulp Pap Can 83:61–66
Shintani N, Sugano Y, Shoda M (2002) Decolourization of Kraft pulp bleaching effluent by a newly isolated fungus, Geotrichum candidum Dec 1. J Wood Sci 48(5):402–408
Sierra-Alvarez R, Lettinga G (1990) The methanogenic toxicity of wood resin constituents. Biol Wastes 33:211–226
Sierra-Alvarez R, Lettinga G (1991) The methanogenic toxicity of wastewater lignins and lignin related compounds. J Chem Technol Biotechnol 50:443–455
Sierra-Alvarez R, Kortekaas S, vanEckort M, Harbrecht J, Lettinga G (1991) The anaerobic biodegradability and methanogenic toxicity of pulping wastewaters. Wat Sci Tech 24:113–125
Singh P (2007) Sequential anaerobic and aerobic treatment of pulp and paper mill effluent in pilot scale bioreactor. J Environ Biol 28(1):77–82
Singh P, Thakur IS (2006) Colour removal of anaerobically treated pulp and paper mill effluent by microorganisms in two steps bioreactor. Bioresour Technol 97(2):218–223
Singhal A, Thakur J (2009) Decolourization and detoxification of pulp and paper mill effluent by Emericella nidulans var nidulans. Hazard Mater 171(1–3):619–625

Singhal V, Kumar A, Rai JP (2005) Bioremediation of pulp and paper mill effluent with Phanerochaete chrysosporium. J Environ Biol 26(3):525–529

Smith JE, Frailey MM (1990) On site evaluation of a teflon based UV light system and hydrogen peroxide for degradation of colour and chlorinated organics in pine from Kraft mill bleach plant effluent. In: Proceedings of Tappi environmental conference, pp 101–110

Smook GA (1992) Handbook for pulp and paper technologists, 2nd edn. Angus Wilde Publications, Vancouver 419 pp

Sorce Inc. (2003) Fungal/bacterium sequencing program for lignin degradation and colour removal from wastewater at a southeastern Kraft pulp mill. Personal communication with Mike Bowling

Springer AM, Hand VC, Jarvis TS (1994) Electrochemical removal of colour and toxicity from bleached Kraft effluent. In: Proceedings of Tappi international environmental conference, pp 271–279

Stephenson R, Duff S (1996) Coagulation and precipitation of mechanical pulping effluents-I: removal of carbon, colour and turbidity. Wat Res 30(4):781–792

Stuthridge TR, Mcfarlane PN (1994) Adsorbable organic halide removal mechanisms in a pulp and paper mill aerated lagoon treatment system. Water Sci Technol 29(5–6):195–208

Stuthridge TR, Campin DN, Langdon AG, Mackie KL, Mcfarlange PN, Wikins AL (1991) Treatability of bleached Kraft pulp and paper mill wastewaters in a New Zealand aerated lagoon treatment system. Water Sci Technol 24(3/4):309–317

Sullivan EC (1986) The use of advanced treatment methods for removal of color and dissolved solids from pulp and paper wastewater-Master's thesis. Virginia Polytechnic Institute and State University

Sun YB, Guo HY, Joyce TW, Chang HM (1992) A study on the reduction of chlorinated organics in bleach plant effluent by oxidation with oxygen. J Pulp Paper Sci 18(2):J49–J55

Sundman G, Kirk TK, Chang HM (1981) Fungal decolourization of Kraft bleach plant effluent: fate of the chromopheric material. Tappi J 64(9):145–148

Tiku DK, Kumar A, Chaturvedi R, Makhijani SD, Manoharan A, Kumar R (2010) Holistic bioremediation of pulp mill effluents using autochthonous bacteria. Int Biodeterior Biodegradation 64(3):173–183

Tomar P, Allen DG (1991) Removal of organochlorines from Kraft mill effluents by an aerated lagoon. Water Poll Res J Can 26(1):101–104

Tong Z, Wada S, Takao Y, Yamagishi T, Hiroyasu I, Tamatsu K (1999) Treatment of bleaching wastewater from pulp-paper plants in China using enzymes and coagulants. J Environ Sci 11(4):480–484

Tono T, Tani Y, Ono KJ (1968) Microbial treatment of agricultural industrial waste. Part 1: Adsorptions of lignins and clarification of lignin containing liquor by mold. Ferment Technol 46:569–574

Toor A, Singh V, Jotshi CK, Bajpai PK, Verma A (2007) Treatment of bleaching effluent from the pulp and paper industry by photo-catalytic oxidation. Tappi J 6(6):9–13

Torrades F, Peral J, Pérez M, Domènech X, Garcia Hortal JA, Riva MC (2001) Removal of organic contaminants in bleached Kraft effluents using heterogeneous photocatalysis and ozone. Tappi J 84(6):1–10

Turk S (1988) Option for treatment of CTMP effluents. Report WTC Bio-7-1988. Environment Canada, Ottawa

Valenzuela J, Bumann U, Cespedes R, Padilla L, Gonzalez B (1997) Degradation of chlorophenols by Alcaligenes eutrophus TMP 134 (p JP4) in bleached Kraft mill effluent. Appl Environ Microbiol 63(1):227–232

van Driessel B, Christov L (2001) Decolourization of bleach plant effluent by Mucoralean and white-rot fungi in a rotating biological contactor reactor. J Biosci Bioeng 92(3):271–276

Viikari L, Suurnakki A, Gronqvist S, Raaska L, Ragauskas A (2009) Forest products: biotechnology in pulp and paper processing encyclopedia of microbiology, 3rd edn. Elsevier, Amsterdam, pp 80–94

Voss RH (1983) Chlorinated neutral organics in biologically treated bleached Kraft mill effluents. Env Sci Tech 17(9):530–537

Wagner M, Nicell JA (2001) Treatment of a foul condensate from Kraft pulping with horseradish peroxidase and hydrogen peroxide. Water Sci Technol 35(2):485–495

Wang I-C, Pan T-T (1999) Interference of some papermaking chemical additives in the coagulation of wastewater. Taiwan J For Sci 14(4):367–384

Welander T, Anderson PE (1985) Anaerobic treatment of wastewater from the production of chemithermomechanical pulp. Wat Sci Tech 17:103–107

Welander T, Lofqvist A, Selmer A (1997) Upgrading aerated lagoons at pulp and paper mills. Water Sci Technol 35(2–3):117–122

Wilson DG, Holloran MF (1992) Decrease of AOX with various external effluent treatments. Pulp Paper Can 93(12):T372–T378

Wilson RW, Murphy KL, Frenelte EG (1987) Aerobic and anaerobic pretreatment of NSSC and CTMP effluent. Pulp Pap Can 88:T4–T8

Wu J, Xiao Y-Z, Yu H-Q (2005) Degradation of lignin in pulp mill wastewaters by white rot fungi on biofilm. Bioresour Technol 96:1357–1363

Xu M, Wang Q, Hao Y (2007) Removal of organic carbon from wastepaper pulp effluent by lab-scale photo-Fenton process. J Hazard Mater 148(1–2):103–109

Yao WY, Kennedy KJ, Tamchung MH, John D (1994) Pre-treatment of Kraft pulp bleach plant effluents by selected UF membranes. Can J Chem Eng 72(6):991–999

Yeber MC, Rodriquez J, Freer J, Baeza J, Duran N, Mansilla HD (1999) Advanced oxidation of a pulp mill bleaching wastewater. Chemosphere 39(10):1679–1688

Yin CF (1989) Characterization, bacterial and fungal degradation, dechlorination and decolourization of chlorolignins in bleaching effluents. PhD Thesis, North Carolina State University, Raleigh, NC

Yin CF, Joyce TW, Chang HM (1989) Kinetics of bleach plant effluent decolourization by *Phanerochaete chrysosporium*. J Biotechnol 10:67–76

Yin CF, Joyce TW, Chang HM (1990) Dechlorination of conventional softwood bleaching effluent by sequential biological treatment. In: Kirk TK, Chang HM (eds) Biotechnology in pulp and paper manufacture, Butterworth-Heinemann, Newton, Mass, pp 231–234

Young RA, Akhtar M (ed) (1998) Environmentally friendly technologies for the pulp and paper industry. Wiley, New York, ISBN 0-471-15770-8

Yu P, Welander T (1988) Anaerobic toxicity of Kraft bleach effluent. In: Tilche A, Rozzi A (eds) Proceedings of 5th international symposium on anaerobic digestion bologna, Itlay, pp 865–867

Zheng Y, Allen DG (1997) Effect of prehydrolysis of D-stage filtrate on the biotreatability of chlorinated organic compounds in bleached Kraft effluent. Water Res 31:1595–1600

Chapter 6
Conclusions and Future Perspectives

Internal process change is one of the options adopted by the pulp and paper industry to reduce pollution at the source. Studies have shown that large number of pulp and paper mills in the developing countries have significantly reduced effluent discharge even when the production has been increased. Among the various treatment processes currently used for pulp and paper effluent treatment, only a few are commonly adopted especially for tertiary treatment. Some of the treatment processes such as ozonation, Fenton's reagent, adsorption, and membrane technology are efficient but are more expensive. Sedimentation is the most commonly adopted process by the pulp and paper industry to remove suspended solids. Flotation is also commonly used in the pulp and paper industry, but most frequently as a tertiary treatment. Coagulants are a preferred option for removing turbidity and color from the wastewater. Reported results have shown that they are also capable in reducing COD, TOC, and AOX to some extent. Among the coagulants, modified chitosan showed the highest performance for color and TOC removal. Polyelectrolytes are better than alum and they produce less sludge and pose fewer problems with sludge dewaterability than alum. Adsorption processes are useful to remove color, COD, and AOX. They are rather expensive and the pulp and paper industry is employing them widely. Chemical oxidants such as ozone + photocatalysis, and ozone + UV are reported to be efficient in removing COD and TOC and color. Ozone is not commonly used in most countries, not even in Europe but it is emerging in North America. Membrane processes are efficient but fouling of membranes is a problem. In secondary treatment processes, activated sludge is the most commonly used process. Aerated lagoons are efficient in removing BOD, COD, and AOX. Anaerobic contact reactors, anaerobic filters and fluidized bed reactors are suitable in reducing organic pollutants. White-rot fungi—*Phanerochaete chrysosporium* and *Coriolus versicolor* are suitable for efficient degradation of the refractory material. High removals are achieved in the case of the combination of two or more physicochemical processes or combination of physicochemical and biological processes. The confirmation of the reported results, their applicability in the real field, and economic evaluations are

P. Bajpai, *Bleach Plant Effluents from the Pulp and Paper Industry*,
SpringerBriefs in Applied Sciences and Technology,
DOI: 10.1007/978-3-319-00545-4_6, © The Author(s) 2013

very important in adopting the process. One of the drawbacks associated with the fungal treatment has been the requirement of an easily metabolizable cosubstrate like glucose for the growth and development of ligninolytic activity. To make the fungal treatment method economically feasible, there is a need to reduce the requirement of cosubstrate or identify a cheaper cosubstrate. Hence, efforts should be made to identify the strains that show good decolorization with less or no cosubstrate and can utilize industrial waste as a cosubstrate. Efforts should be also made to utilize the spent fungal biomass for preparing the culture medium required in the synthesis of active fungal biomass. If successful, the cost of treatment may be further reduced. Since a lignin degrading system of white-rot fungus has a high oxygen requirement, use of oxygen instead of air as fluidizing media should be explored. Increasing the oxygen concentration in the culture atmosphere is expected to have a dual effect: it would lead to an increased titer of the lignin degrading system and to increased stability of the existing system. A quantitative study of extracellular enzymes is also required in order to gain insight into the possible enzymatic mechanism involved in the degradation of lignin-derived compounds present in the effluents.

Use of white-rot fungi as a pretreatment prior to bacterial treatment to enhance the bacterial ability to remove organic chlorine and to degrade the relatively higher molecular weight chlorolignins is an attractive option. This process can be used as an alternative to internal process modifications (modified cooking, oxygen bleaching, high level chlorine dioxide substitution etc.), and conventional biological treatment. Since the most AOX and color is found in high molecular weight chlorolignins, research should concentrate on the fate of high molecular weight chlorolignins in biological treatment and in the natural environment. Since bacteria degrade significantly only those chloroorganics with molecular weights lower than 800–1,000 Da, research is needed to decrease the chlorolignin molecular weight or to remove high molecular weight chlorolignins before biological treatment is applied in order to enhance the biotreatability of bleaching effluents.

Glossary

Agro Agricultural materials

AOX Adsorbable Organic Halides

AOP Advanced Oxidation Process

ASB Aerated lagoon treatment

AST Activated Sludge Treatment

BFD Black Furnace Dust

BKME Bleached Kraft Mill Effluent

BOD Biological Oxygen Demand

CEHDED Chlorination-Extraction-Hypo-Chlorine dioxide-Extraction- Chlorine Dioxide

CMP Chemimechanical Pulp

COD Chemical Oxygen Demand, bioCOD, biodegradable COD

CTMP Chemithermomechanical Pulp

DOC Dissolved Organic Carbon

ECF Elemental Chlorine Free

EDTA Ethylenediaminetetraacetic acid

EO Extraction with Oxygen

EOP Extraction with Oxygen and Peroxide

EOX Extractable organic halide

FPL Forest product laboratory

Floobed Floating biological bed

P. Bajpai, *Bleach Plant Effluents from the Pulp and Paper Industry*,
SpringerBriefs in Applied Sciences and Technology,
DOI: 10.1007/978-3-319-00545-4, © The Author(s) 2013

GMF Granular Membrane Filtration

HRT Hydraulic Retention Time

KL Kraft Lignin

LSHS Low Sulphur Heavy Stock

MF Membrane Filtration

MFO Mixed Function Oxygenase

MyCoR Mycelia Colour Removal

NCSU North Carolina State University

NSSC Neutral Sulfite Semi-Chemical pulping

PARCOM Paris Convention for the Prevention of Marine Pollution

PGW Pressure Groundwood Pulping

PHT Pressurized hydrogen peroxide treatment

PO Pressurized oxygen

RBC Rotating Biological Contractor

RFA Resin and Fatty Acids

SBR Sequencing Batch Reactors

SGW Stone Groundwood Pulping

SRT Solid Retention Time

TCF Total Chlorine-Free

TDS Total dissolved solids

TMP Thermo-Mechanical Pulping

TOC Total Organic Carbon

TOCl Total Organic Chlorine

TSS Total Suspended Solids

UASB Upflow Anaerobic Sludge Blankets

UF Ultrafiltration

UV Ultraviolet

VP Versatile Peroxidase

Index